NANOMECHANICS AND MICROMECHANICS

Generalized Models and Nonclassical Engineering Approaches

AAP Research Notes on Nanoscience and Nanotechnology

NANOMECHANICS AND MICROMECHANICS

Generalized Models and Nonclassical Engineering Approaches

Edited by

Satya Bir Singh, PhD

Alexander V. Vakhrushev, DSc

A. K. Haghi, PhD

AAP | APPLE ACADEMIC PRESS

Apple Academic Press Inc.
4164 Lakeshore Road
Burlington ON L7L 1A4, Canada

Apple Academic Press Inc.
1265 Goldenrod Circle NE
Palm Bay, Florida 32905, USA

Exclusive worldwide distribution by CRC Press, a member of Taylor & Francis Group

No claim to original U.S. Government works

International Standard Book Number-13: 978-1-77188-833-2 (Hardcover)
International Standard Book Number-13: 978-0-42932-244-0 (eBook)

Library and Archives Canada Cataloguing in Publication

Title: Nanomechanics and micromechanics : generalized models and nonclassical engineering approaches / edited by Satya Bir Singh, PhD, Alexander V. Vakhrushev, DSc, A.K. Haghi, PhD.

Names: Singh, Satya Bir, editor. | Vakhrushev, Alexander V., editor. | Haghi, A. K., editor.

Series: AAP research notes on nanoscience & nanotechnology.

Description: Series statement: AAP research notes on nanoscience & nanotechnology | Includes bibliographical references and index.

Identifiers: Canadiana (print) 2020018511X | Canadiana (ebook) 20200185187 | ISBN 9781771888332 (hardcover) | ISBN 9780429322440 (ebook)

Subjects: LCSH: Micromechanics—Mathematical models. | LCSH: Nanoelectromechanical systems—Mathematical models.

Classification: LCC QC176.8.M5 N36 2020 | DDC 620.1/1299—dc23

Library of Congress Cataloging-in-Publication Data

Names: Singh, Satya Bir, editor. | Vakhrushev, Alexander V., editor. | Haghi, A. K., editor.

Title: Nanomechanics and micromechanics : generalized models and nonclassical engineering approaches / edited by Satya Bir Singh, Alexander V. Vakhrushev, A.K. Haghi.

Other titles: AAP research notes on nanoscience & nanotechnology.

Description: Burlington ON, Canada ; Palm Bay, Florida : Apple Academic Press, [2020] | Series: AAP research notes on nanoscience and nanotechnology | Includes bibliographical references and index. | Summary: "This volume, Nanomechanics and Micromechanics: Generalized Models and Nonclassical Engineering Approaches, enables readers to interpret and predict the effective mechanical properties of existing and emerging composites through modeling and design. The book addresses that materials and structures with small-scale dimensions do not behave in the same manner as their bulk counterparts. Once materials dimensions are reduced to the micron- and-sub-micron range, their properties are subject to significant change. Thus, mechanical properties will be varied and will depend on the sample size. In the meantime, due to the large surface-to-volume ration of small structures, deformation mechanisms are subject to change. This volume integrates various approaches in micromechanics and nanomechanics into a unified mathematical framework, complete with coverage of both linear and nonlinear behaviors. It weaves together the basic concepts, mathematical fundamentals, and formulations of micromechanics and nanomechanics into a systemic approach for understanding and modeling the effective material behavior of composite materials. While providing information on recent developments in the mathematical framework of micro- and nanomechanics, the volume addresses highly localized phenomena and a number of interesting applications. It also illustrates application of micromechanical and nanomechanical theory to design novel engineering materials. This volume is intended for advanced undergraduate and graduate students, researchers, and engineers interested and involved in mechanical analysis and design. Key features: Covers some recent applications and impact areas of micromechanics and nanomechanics that have not been discussed in traditional micromechanics and nanomechanics books Reviews the fundamentals for micro- and nanomechanics Describes nano and microstructure characterization Presents case studies and research methodology on nanomechanics and micromechanics Studies mathematical models and failure criteria Introduces a straightforward approach on derivation of mechanical/mathematical results with emphasis on issues of practical importance"-- Provided by publisher.

Identifiers: LCCN 2020007083 (print) | LCCN 2020007084 (ebook) | ISBN 9781771888332 (hardcover) | ISBN 9780429322440 (ebook)

Subjects: MESH: Nanotechnology | Microtechnology

Classification: LCC R857.N34 (print) | LCC R857.N34 (ebook) | NLM QT 36.5 | DDC 610.28--dc23

LC record available at https://lccn.loc.gov/2020007083

LC ebook record available at https://lccn.loc.gov/2020007084

Apple Academic Press also publishes its books in a variety of electronic formats. Some content that appears in print may not be available in electronic format. For information about Apple Academic Press products, visit our website at **www.appleacademicpress.com** and the CRC Press website at **www.crcpress.com**

ABOUT THE AAP RESEARCH NOTES ON NANOSCIENCE & NANOTECHNOLOGY BOOK SERIES:

AAP Research Notes on Nanoscience & Nanotechnology reports on research development in the field of nanoscience and nanotechnology for academic institutes and industrial sectors interested in advanced research.

BOOKS IN THE AAP RESEARCH NOTES ON NANOSCIENCE & NANOTECHNOLOGY BOOK SERIES:

- **Nanostructure, Nanosystems and Nanostructured Materials: Theory, Production, and Development**
  Editors: P. M. Sivakumar, PhD, Vladimir I. Kodolov, DSc, Gennady E. Zaikov, DSc, A. K. Haghi, PhD
- **Nanostructures, Nanomaterials, and Nanotechnologies to Nanoindustry**
  Editors: Vladimir I. Kodolov, DSc, Gennady E. Zaikov, DSc, and A. K. Haghi, PhD
- **Foundations of Nanotechnology: Volume 1: Pore Size in Carbon-Based Nano-Adsorbents**
  A. K. Haghi, PhD, Sabu Thomas, PhD, and Moein MehdiPour MirMahaleh
- **Foundations of Nanotechnology: Volume 2: Nanoelements Formation and Interaction**
  Sabu Thomas, PhD, Saeedeh Rafiei, Shima Maghsoodlou, and Arezo Afzali
- **Foundations of Nanotechnology: Volume 3: Mechanics of Carbon Nanotubes**
  Saeedeh Rafiei
- **Engineered Carbon Nanotubes and Nanofibrous Material: Integrating Theory and Technique**
  Editors: A. K. Haghi, PhD, Praveen K. M., and Sabu Thomas, PhD
- **Carbon Nanotubes and Nanoparticles: Current and Potential Applications**
  Editors: Alexander V. Vakhrushev, DSc, V. I. Kodolov, DSc, A. K. Haghi, PhD, and Suresh C. Ameta, PhD
- **Advances in Nanotechnology and the Environmental Sciences: Applications, Innovations, and Visions for the Future**
  Editors: Alexander V. Vakhrushev, DSc, Suresh C. Ameta, PhD, Heru Susanto, PhD, and A. K. Haghi, PhD
- **Chemical Nanoscience and Nanotechnology: New Materials and Modern Techniques**
  Editors: Francisco Torrens, PhD, A. K. Haghi, PhD, and Tanmoy Chakraborty, PhD
- **Nanomechanics and Micromechanics: Generalized Models and Nonclassical Engineering Approaches**
  Editors: Satya Bir Singh, PhD, Alexander V. Vakhrushev, DSc, and A. K. Haghi, PhD

ABOUT THE EDITORS

Satya Bir Singh, PhD
Professor, Department of Mathematics, Punjabi University, Patiala, India

Satya Bir Singh, PhD, is a Professor of Mathematics at Punjabi University Patiala in India. Prior to this he has worked as an Assistant Professor in Mathematics in the Thapar Institute of Engineering and Technology, Patiala, India. He has published about 125 research papers in journals of national and international repute and has given invited talks at various conferences and workshops and has also organized several national and international conferences. He has been a Coordinator and Principal Investigator of several schemes funded by the Department of Science and Technology, Government of India, New Delhi; the University Grants Commission, Government of India, New Delhi; and the All India Council for Technical Education, Government of India, New Delhi. He has 21 years of teaching and research experience. His areas of interest include mechanics of composite materials, optimization techniques, and numerical analysis. He is a life member of various learned bodies.

Alexander V. Vakhrushev, DSc
Professor, M.T. Kalashnikov Izhevsk State Technical University, Izhevsk, Russia; Head, Department of Nanotechnology, and Microsystems of Kalashnikov Izhevsk State Technical University, Russia

Alexander V. Vakhrushev, DSc, is Professor at the M.T. Kalashnikov Izhevsk State Technical University in Izhevsk, Russia, where he teaches theory, calculating, and design of nano- and microsystems. He is also the Chief Researcher of the Department of Information-Measuring Systems of the Institute of Mechanics of the Ural Branch of the Russian Academy of Sciences and Head of the Department of Nanotechnology and Microsystems of Kalashnikov Izhevsk State Technical University. He is a Corresponding Member of the Russian Engineering Academy. He has over 400 publications to his name, including monographs, articles, reports, reviews, and patents. He has received several awards, including an Academician A. F. Sidorov Prize from the Ural Division of the Russian Academy of Sciences for significant contribution to the creation of the theoretical fundamentals of physical processes taking place in multi-level nanosystems and was proclaimed an Honorable Scientist of the Udmurt Republic. He is currently a member of editorial boards of several journals, including *Computational Continuum Mechanics, Chemical Physics and Mesoscopia*, and *Nanobuild*. His research interests include multiscale mathematical modeling of physical-chemical processes into the nano-hetero systems at nano-, micro- and macro-levels; static and dynamic interaction of nanoelements; and basic laws relating the structure and macro characteristics of nano-hetero structures.

A. K. Haghi, PhD
Editor-in-Chief, International Journal of Chemoinformatics and Chemical Engineering and Polymers Research Journal; Member, Canadian Research and Development Center of Sciences and Cultures (CRDCSC), Canada

A. K. Haghi, PhD, is the author and editor of 165 books as well as of 1000 published papers in various journals and conference proceedings. Dr. Haghi has received several grants, consulted for a number of major corporations, and is a frequent speaker to national and international audiences. Since 1983, he served as professor at several universities. He is currently Editor-in-Chief of the *International Journal of Chemoinformatics and Chemical Engineering* and the *Polymers Research Journal* and is on the editorial boards of many international journals. He is also a member of the Canadian Research and Development Center of Sciences and Cultures, Montreal, Quebec, Canada. He holds a BSc in urban and environmental engineering from the University of North Carolina (USA); an MSc in mechanical engineering from North Carolina A&T State University (USA); a DEA in applied mechanics, acoustics, and materials from the Université de Technologie de Compiègne (France); and a PhD in engineering sciences from the Université de Franche-Comté (France).

CONTENTS

CONTRIBUTORS

A. Yu. Fedotov
Department of Mechanics of Nanostructures, Institute of Mechanics,
Udmurt Federal Research Center, Ural Division, Russian Academy of Sciences, Izhevsk, Russia;
Department of Nanotechnology and Microsystems, Technic Kalashnikov Izhevsk State Technical University, Izhevsk, Russia, E-mail: alezfed@gmail.com

Nishi Gupta
Department of Mathematics, UIS, Chandigarh University, Gharuan, India

Vandana Gupta
Department of Mathematics, Dashmesh Khalsa College, Zirakpur (Mohali), Punjab – 160059, India

V. I. Kodolov
Basic Research – High Educational Center of Chemical Physics and Mesoscopics, Studencheskaya str. 7, 423067, Izhevsk, Russia | M.T. Kalashnikov Izhevsk State Technical University, Studencheskaya str. 7, 423067, Izhevsk, Russia, E-mail: vkodol.av@mail.ru

Fang-Yie Leu
Computer Science Department, Tunghai University, Taiwan

G. Rathnakar
Department of Mechanical Engineering, ATME College of Engineering, Bannur Road, Mysore, Karnataka – 570028, India

G. Ravichandran
Assistant Professor, Department of Mechanical and Automobile Engineering,
Christ (Deemed-to-be-University), Kanminike, Kumbalgodu, Mysore Road, Kengeri, Bengaluru – 560074, Karnataka, India, E-mail: ravichandran.g@chirstuniversity.in |
Research Scholar, VTU-RRC, Department of Mechanical Engineering, Belagavi, Karnataka – 590018, India

N. Santhosh
Assistant Professor, Department of Mechanical and Automobile Engineering,
Faculty of Engineering, Christ (Deemed-to-be-University), Kanminike, Kumbalgodu, Mysore Road, Kengeri, Bengaluru – 560074, Karnataka, India, E-mail: santhosh.n@chirstuniversity.in

I. N. Shabanova
Basic Research – High Educational Center of Chemical Physics and Mesoscopics, Studencheskaya str. 7, 423067, Izhevsk, Russia; Udmurt Federal Research Center, Ural Division, RAS, T. Baramzina str., 34, 423066, Izhevsk, Russia

Shivdev Shahi
Department of Mathematics, Punjabi University Patiala, Punjab – 147002, India, E-mail: shivdevshahi93@gmail.com

D. S. Shuklin
Kalashnikov Izhevsk State Technical University, Izhevsk, Russia

S. G. Shuklin
Kalashnikov Izhevsk State Technical University, Izhevsk, Russia | Udmurt State University, Izhevsk, Russia, E-mail: shuklin_sg@mail.ru

Satya Bir Singh
Department of Mathematics, Punjabi University Patiala, Punjab – 147002, India, E-mail: sbsingh69@yahoo.com

Heru Susanto
School of Business and Economics, University Brunei Darussalam, Brunei, E-mail: heru.susanto@lipi.go.id

S. V. Suvorov
Department of Mechanics of Nanostructures, Institute of Mechanics, Udmurt Federal Research Center, Ural Division, Russian Academy of Sciences, Izhevsk, Russia

N. S. Terebova
Basic Research – High Educational Center of Chemical Physics and Mesoscopics, Studencheskaya str. 7, 423067, Izhevsk, Russia | Udmurt Federal Research Center, Ural Division, RAS, T. Baramzina str. 34, 423066, Izhevsk, Russia

Pankaj Thakur
Department of Mathematics, ICFAI University, Solan, Himachal Pradesh – 174103, India

V. V. Trineeva
Basic Research – High Educational Center of Chemical Physics and Mesoscopics, Studencheskaya str. 7, 423067, Izhevsk, Russia | Udmurt Federal Research Center, Ural Division, RAS, T. Baramzina str. 34, 423066, Izhevsk, Russia

Alexander V. Vakhrushev
Kalashnikov Izhevsk State Technical University, Izhevsk, Russia | Institute of Mechanics, Ural Division, Russian Academy of Sciences, Izhevsk, Russia, E-mail: vakhrushev-a@yandex.ru

L. Francis Xavier
Assistant Professor, Department of Mechanical and Automobile Engineering, Christ (Deemed-to-be-University), Kanminike, Kumbalgodu, Mysore Road, Kengeri, Bengaluru – 560074, Karnataka, India, E-mail: francis.xavier@chirstuniversity.in

ABBREVIATIONS

AMMC	aluminum metal matrix composite
ANOVA	analysis of variance
APPh: Cu/CNC	ammonium polyphosphate to the nanogranul of copper/carbon nanocomposite
ASTM	American Society for Testing and Materials
BDMA	dimethyl benzylamine
C	coke
CL	cellulose
CMS	course management systems
Cu/CNC	copper/carbon nanocomposite
DBMSs	database management system software
DFT	density functional theory
DGEBA	diglycidyl ether bisphenol
EDAX	energy dispersive x-ray spectroscopy
EMIS	education management information system
EPR	electron paramagnetic resonance
ES	epoxy resin
FGM	functionally graded material disc
FRS	fire retardant systems
GIS	geographic information system
HNTs	halloysite nanotubes
IS	information system
LMS	learning management systems
MAF	monoammonium phosphate
MEAM	modified embedded-atom method
Mg	magnesium
MML	mechanically mixed layer
NESIS	National Education Statistical Information Systems
Ni/CNC	nickel/carbon nanocomposite
PA	polyamide
PAN	polyacrylonitrile
Pb	lead
PC	polycarbonate

PE	polyethylene
PET	polyethylene terephthalate
PG	particle gradient
PP	polypropylene
PS	polystyrene
PU	polyurethane
R	ratios
RT	room temperature
TEM	transition electron microscopy
TETA	triethylenetetramine
TF	triaxiality factor
UCAS	Universities and Colleges Admission Service
VLE	virtual learning environments

SYMBOLS

$\Delta\sigma_y$	tensile residual stress
ΔH_b	heat of boiling
ΔH_m	heat of fusion
H	internal energy
$Re \to 0$	Reynolds number tends to zero
T_b	boiling point
Tg	vitrification temperature
T_m	melting temperature

PREFACE

Materials and structures with small-scale dimensions do not behave in the same manner as their bulk counterparts.

Once materials dimensions are reduced to the micron-and-sub-micron range, their properties are subject to significant change. Thus, mechanical properties will be varied and will depend on the sample size. In the meantime, due to the large surface-to-volume ration of small structures, deformation mechanisms are subject to change.

As structures move beyond the submicrometer to the nanometer scale, the description of mechanical behavior focuses on issues other than the traditional ensemble of defects. For instance, the length scale of a typical dislocation and the volume of material required for it to have a significant influence on deformation are large compared to the typical volume of a nanosized object.

This volume enables readers to interpret and predict the effective mechanical properties of existing and emerging composites through modeling and design.

This book integrates various approaches in micromechanics and nano-mechanics into a unified mathematical framework complete with coverage of both linear and nonlinear behaviors.

It weaves together the basic concepts, mathematical fundamentals, and formulations of micromechanics and nanomechanics into a systemic approach for understanding and modeling the effective material behavior of composite materials.

This volume is intended for advanced undergraduate and graduate students, researchers and engineers interested and involved in mechanical analysis and design. It also serves as a research-oriented reference book, intended for first-year postgraduate students in materials science, applied computational mechanics, nano-science and technology, and mechanical engineering. It is ideal as a reference book for graduate students with an interest in the computational nano and micromechanical analysis of new materials and presents a broad exposition of analytical and mathematical methods for modeling engineering materials.

While providing recent developments in the mathematical framework of micro- and nanomechanics, it addresses highly localized phenomena and a number of interesting applications.

It also illustrates the application of micromechanical and nanomechanical theory to design novel engineering materials.

This book:

- covers some recent applications and impact areas of micromechanics and nanomechanics that have not been discussed in traditional micromechanics and nanomechanics books;
- reviews fundamentals for micro and nanomechanics;
- describes nano and microstructure characterization;
- presents case studies and research methodology on nanomechanics and micromechanics;
- studies mathematical models and failure criteria; and
- introduces a straightforward approach to the derivation of mechanical/mathematical results with emphasis on issues of practical importance.

CHAPTER 1

EFFECTS OF HEAT TREATMENT CONDITIONS ON MICROSTRUCTURE AND MECHANICAL PROPERTIES OF HALLOYSITE NANOTUBE (HNT) FILLED EPOXY NANOCOMPOSITES

G. RAVICHANDRAN,[1,3] G. RATHNAKAR,[2] and N. SANTHOSH[3]

[1]*Research Scholar, VTU-RRC, Department of Mechanical Engineering, Belagavi, Karnataka – 590018, India*

[2]*Department of Mechanical Engineering, ATME College of Engineering, Bannur Road, Mysore, Karnataka – 570028, India*

[3]*Department of Mechanical Engineering, Faculty of Engineering, Christ (Deemed to be University), Mysore Road, Bengaluru, Karnataka – 560074, India*

ABSTRACT

Halloysite nanotubes (HNTs) are naturally occurring Kaolite group minerals having an aluminosilicate-layer in the form of nanotubes. The composites containing a nanofiller in the form of HNTs used as a reinforcement and an epoxy resin as a matrix is effectively fabricated by using polymer casting method incorporating dispersion strengthening by mechanical stirring action and specimens are prepared as per American Society for Testing and Materials (ASTM) standards. The functional nanofillers are effectively used to enhance the mechanical strength of the composite by restricting matrix disengagement movements. In this regard, the present work is carried out to evaluate the mechanical properties of the nanocomposites consisting of different Wt.% of HNTs effectively heat-treated at three temperatures viz. Room temperature (RT), 50°C and 70°C and varying in the range of 0 to 10 with an interval of 5. The various properties viz. "density, hardness, tensile, flexural, and impact strength" are investigated through the ASTM procedure. As per the experimental investigation, the mechanical properties of the nanocomposite increases by the incorporation of heat-treated HNT. Further, the study revealed that the 10 Wt.% of HNT with 50°C heat-treated nanocomposite shows superior tensile and flexural strength. However, the critical observation of the results reveal that the impact strength is maximum for 70°C heat-treated nanocomposite synthesized for 5 Wt.% of HNT. It is evident that the properties of the nanocomposite depend on the quantity of functional filler present and temperature of heat treatment.

1.1 INTRODUCTION

Nanocomposites are materials which take the benefit of dispersoids, whiskers, and platelet form of nanoparticles used as reinforcement in their fabrication. In general, nanofiller with a size range of several hundred nanometers and any polymer matrix phase, either a thermoset or thermoplastic comprises a nanocomposite. Due to the better interfacial area property, the nanoparticles are the most commonly used reinforcement materials compared to other reinforcements [1, 2].

Halloysite nanotube (HNT) is basically a kaolite group naturally occurring mineral which includes aluminosilicate nanotubes in layer forms. HNTs are available in the form of white color nanoparticles with chemical formula $Al_2Si_2O_5$ (OH)4•$2H_2O$. HNTs are odorless and more economical compared to other nanofillers, especially carbon nanotubes [3–6]. Studies revealed that mechanical properties of the polymer matrix can be enhanced by the addition of HNTs, e.g., toughness, and elastic modulus can be enhanced through restrictive matrix dislocation activity [7–10]. The effective dispersion of HNTs in the polymer matrix are known to improve the characteristic features, however, it is very difficult to disperse HNTs effectively into the polymer matrix due to clusterness of the HNTs [11–13]. Therefore, it is a challenge for the researchers to establish the right parameters to achieve a homogenous dispersion of HNTs in the polymer matrix.

Currently, the methods such as mechanical stirring, ultrasonic homogenization, and ball milling are commonly used to disperse nanofillers in the polymer matrix [14]. The mechanical stirring is an easy method to achieve homogenous dispersions with the minimum aggregation of nanofillers in the polymer matrix and also it is more suitable for the production of the nanocomposites commercially [15, 16]. In the literature, it is revealed that the HNTs have an appreciable amount of water between SiO_4 and AlO_6 structures [17–19]. Therefore, the structure, chemical, and mechanical properties of HNT reinforced polymer composites depend majorly upon heat-treated temperatures.

In the present work, a set of experiments is carried out to analyze the dispersion behavior and structural changes in HNTs at different heat-treated temperatures. Further, the mechanical properties of HNT-reinforced epoxy nanocomposites are investigated and from the revelations, it is concluded that the addition of HNT nanofillers with different weight proportions in epoxy improves their mechanical properties than compared to neat composites.

1.2 MATERIALS AND METHODS

1.2.1 MATERIALS AND FABRICATION PROCEDURE

HNTs are used as reinforcements in current work and they are acquired from Sigma Aldrich Company, Bengaluru, India. The diameter and length of the HNT's has a range in between 30 to 70 nm and 1–15 μm, respectively. The morphology of HNT has a tube-like structure with a density of 2.53 g/cc and the surface area is about 65 m^2/g. It has a high aspect ratio and low percolation property which makes it convenient to be used as reinforcement for epoxy matrix composites. The vacuum furnace (Figure 1.1) is used to heat treat the HNTs at temperatures, viz. 50°C and 70°C for 10 minutes and are subsequently cooled down over a duration of 1 hour. Table 1.1 shows the compositions and specifications of the constituents of nanocomposite.

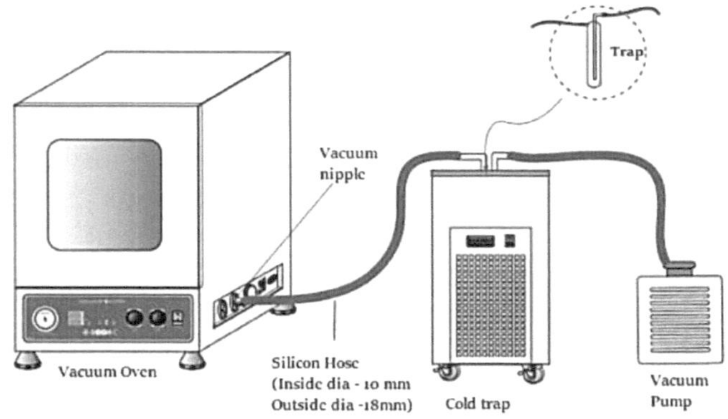

FIGURE 1.1 Vacuum furnace used for heat treatment.

TABLE 1.1 Specification of Materials Used in Nanocomposite

Reinforcement	
Halloysite Nanotube (HNT)	– Sigma-Aldrich – Bengaluru, India – Stock No. NS6130-09-917 – CAS-No.: "1332-58-7" – Formula: "$Al_2Si_2O_5(OH)4 \cdot 2\ H_2O$" – Molecular Weight: "294.19 g/mol" – Purity: 99.9%
Matrix	
Epoxy Resin	– Atul India Ltd., Gujarat, India – Product: Lapox L-12 – Chemical name: Diglycidyl Ether Bisphenol (DGEBA) – Density: 1120 g/cc – Gelation Time at 80°C: 205 minutes – Glass Transition Temperature: 140°C – Curing time: 4 hrs. at 140°C
Hardener	
Tryethylene Tetramine (TETA)	– Atul India Ltd., Gujarat, India – Product: K-6 – Density: 954 g/cc
Accelerator	
N,N- dimethyl benzylamine (BDMA)	– Sigma-Aldrich, Bengaluru, India – CAS-No.: "103-83-3" – Accelerator – Density: "0.9 g/ml at 25°C (lit)"

Heat-treated HNTs considered for present work are thoroughly mixed in the epoxy resin with a hardener in the ratio 100:12 by weight using a mechanical stirrer (as shown in Figure 1.2) maintaining a constant speed of 500 r.p.m., followed by the vacuum cleaning carried out for a duration of 10 min. The mixture is then poured into molds and allowed to cure at RT for 24 hours. Further, post-curing is done at a temperature of 50°C for 2 hours. Three different fractions of HNT (0, 5 and 10 Wt.%) are used to reinforce the epoxy.

There are five samples for each test that are prepared in accordance with the American Society for Testing and Materials (ASTM) standards through the polymer casting method, The specimens fabricated are as shown in Figure 1.3. The details of the constituents included in prepared composites are tabulated in Table 1.2.

TABLE 1.2 Details of the Constituents With Designation for Each of the Sample of the Composites

	Weight Percent		Weight Percent		Weight Percent
Untreated (RT) HNTs	5 10	50°C Heat-treated HNTs	5 10	70°C Heat-treated HNTs	5 10

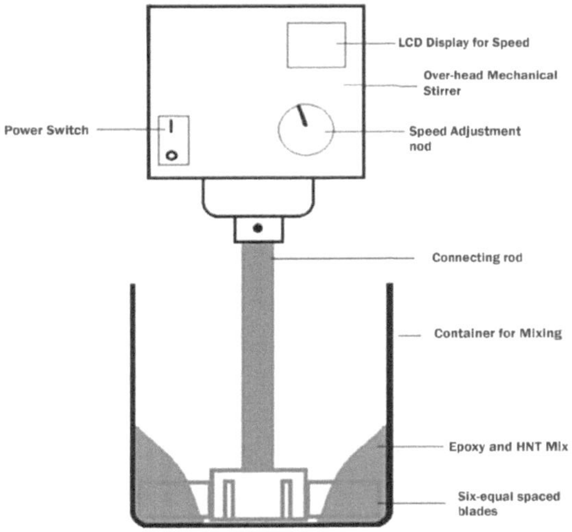

FIGURE 1.2 Mechanical stirrer for dispersion process.

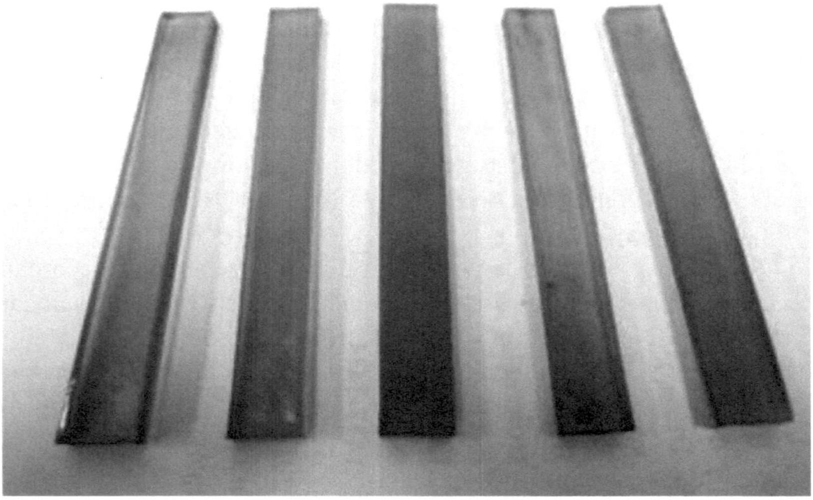

FIGURE 1.3 Specimen as per ASTM standards for Mechanical characterization.

1.2.2 EXPERIMENTATION

1.2.2.1 MICROSCOPIC EXAMINATION

The heat-treated HNT particles are inspected in a transmission electron microscope (TEM) equipped with a computer image analysis system; the TEM measurements are carried out with a 1200 EX TEM applying an accelerating voltage of 120 keV.

1.2.2.2 MECHANICAL PROPERTIES TESTING PROCEDURE

The density of the composites was determined by using a high precision electronic balance (Mettler Toledo, Model AX 205) using Archimede's principle. Hardness (Shore-D) of the samples was measured as per ASTM D2240, by using a Hiroshima make hardness tester (Durometer). Six readings at different locations were noted and the average value is reported. The tensile measurement was carried out using a universal tensile testing machine (JJ Lloyd, London, United Kingdom, capacity 1–20 kN), according to ASTM D3039. The tensile test was performed at a crosshead speed of 30 mm/min (quasi-static).

The flexural strength of the particulate filled composites is determined on rectangular specimens (90 mm × 12 mm × 3 mm) in three points bending at RT according to ASTM D790. The span length of the specimens was 70 mm and their loading on a universal testing machine (JJ Lloyd, London, United Kingdom, capacity 1–20 kN) occurred with deformation rate v = 1.3 mm/min.

Izod impact test was carried out using an Avery ceast pendulum impact tester (ASTM D256–92). A 7.5 J/m energy hammer with the striking velocity is 3.46 m/s is used. Five samples are tested for each composite type for all the studies and the average value is recorded.

1.3 RESULTS AND DISCUSSION

1.3.1 MICROSTRUCTURE OF HEAT-TREATMENT HNTS

TEM images of HNTs reinforced epoxy nanocomposites are presented in Figure 1.4(a), (b), and (c), respectively. In both heat-treated and untreated conditions, the most of HNTs particles are uniformly dispersed at nanometer scale in an epoxy matrix. HNTs are multi-wall nanotubes consisting of aluminosilicate layers, which are curved and closely packed. The outer diameters, inner diameters, and lengths of the nanotubes are about 30–180 nm, 10–30 nm, and 2–10 μm, respectively. The distinctive boundaries between each layer can be clearly visible is as shown in Figure 1.4(a).

The HNTs are heat-treated at temperatures viz. untreated HNTs (at RT), 50°C and 70°C, Figure 1.4(b) shows that HNTs particles are irregular in shape and also formed agglomeration by itself. Further, as the heat treatment temperature increases the structure of the HNTs formed like amorphous, which are even more undefined in shape is as shown in Figure 1.4(c). Therefore, from this study, it was concluded that the hydroxyl groups of HNTs are eliminated due to dehydration as the heat temperature increases. And also, temperature more than 50°C influences extensively on the structural change of the HNTs and predicted that these structural changes should be strongly monitoring the behavior of the HNTs which was used as a reinforcement material for the epoxy matrix.

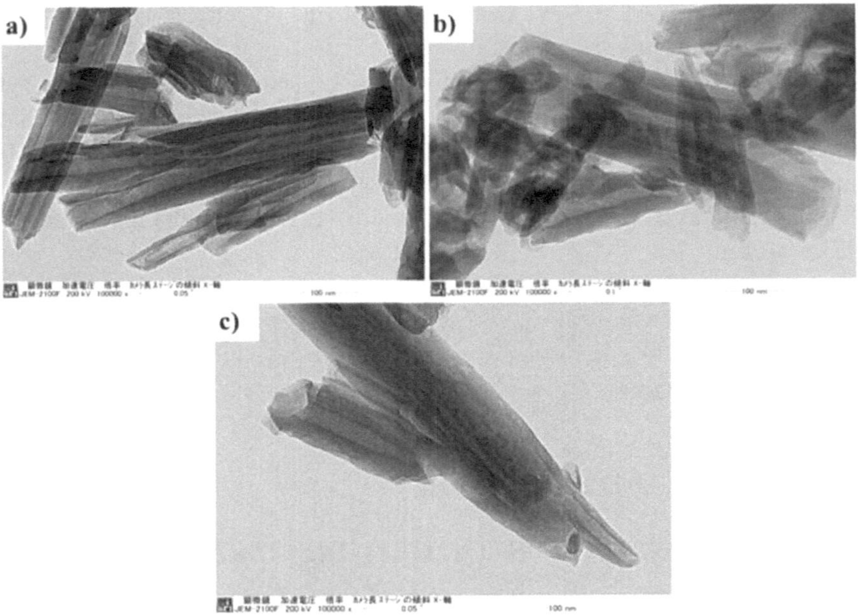

FIGURE 1.4 TEM images of HNT reinforced epoxy nanocomposites. (a) Untreated (RT) with 10 Wt.% HNT, (b) heat-treated at 50°C with 10 Wt.% HNT, (c) heat-treated at 70°C with 10 Wt.% HNT.

1.3.2 DENSITY OF EPOXY/HNT NANOCOMPOSITES

In order to obtain optimized dispersion conditions for improving mechanical properties, the specimens for various tests are synthesized using heat-treated HNTs at various temperatures. The density of a composite depends on the relative proportion of matrix and reinforcing materials. Figure 1.5 shows an increasing trend in density observed with the increasing content of HNT particles in the epoxy matrix. By contrast, a marginal improvement in density is observed in nanocomposites of both untreated and heat-treated HNTs and also the structural changes of HNTs at various temperatures which affect significantly on the density of Epoxy/HNT nanocomposites are inferred. The highest density is observed in the case of 10 Wt.% HNTs-Epoxy system at a heat-treated temperature of 70°C due to better dispersion of the reinforcements as well as enhanced agglomeration.

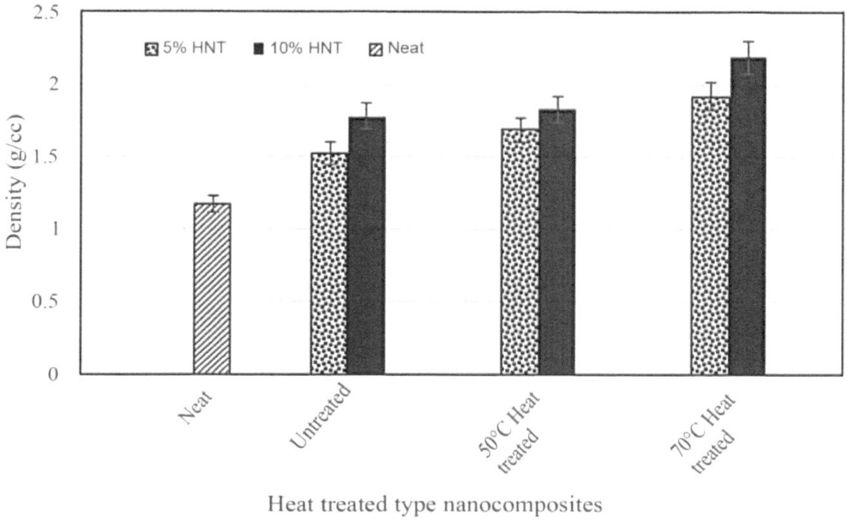

FIGURE 1.5 Density of HNT filled epoxy nanocomposites.

1.3.3 HARDNESS OF EPOXY/HNT NANOCOMPOSITES

The surface hardness is considered to be one of the important factors to be determined and it has a major effect on the wear rate of the composites. The test result shows that there is an increase in the hardness value of the nanocomposites by the incorporation of HNT content. The hard nature of epoxy with exfoliation of polymer chains in between two plates of nano-clay (HNT) makes the surface of composite very hard due to which the indentation of the indenter is quite difficult. The critical observation of the graph as seen in Figure 1.6 reveals that there is no specific trend with variation in heat-treated temperatures. Improvement in hardness is observed only in nanocomposite that contained 50°C heat-treated HNT as the reinforcement and for 70°C heat-treated HNTs, nanocomposites exhibited a decrease in value even with the addition of 10 Wt.% HNT. This signified that the structural changes of HNTs at various temperatures are involved to obtain a formidable chemical combination between epoxy and HNTs.

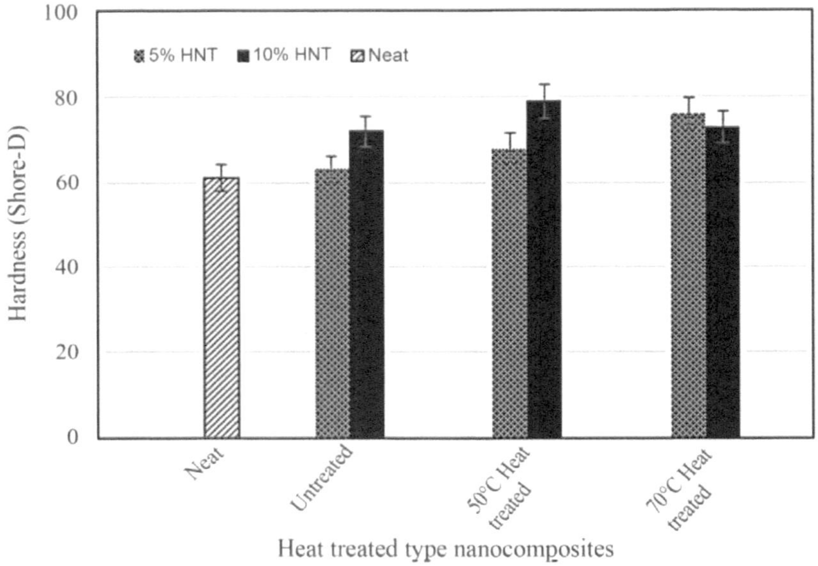

FIGURE 1.6 Hardness of epoxy/HNT nanocomposites.

1.3.4 TENSILE BEHAVIOR OF EPOXY/HNTS NANOCOMPOSITES

From Figure 1.7, it is evident that the tensile strength of composites is found to increase with HNT reinforcement compared to virgin epoxy due to the restriction of the mobility and deformability of the epoxy and also the formation of ordered exfoliation of polymer chains in between the interstitial spacing of nanoclay (HNT). The addition of heat-treated HNT in epoxy increased the interfacial stiffness and static adhesion strength of the composites compared to neat, which constitutes to transfer the elastic deformation to a great extent. The results showed a decrease in tensile strength in 70°C heat-treated HNT due to poor bonding at the interface between epoxy and HNT nanoparticles. This low degree of interfacial interaction will result in a decrease in tensile strength even with the higher concentration of HNT reinforcements.

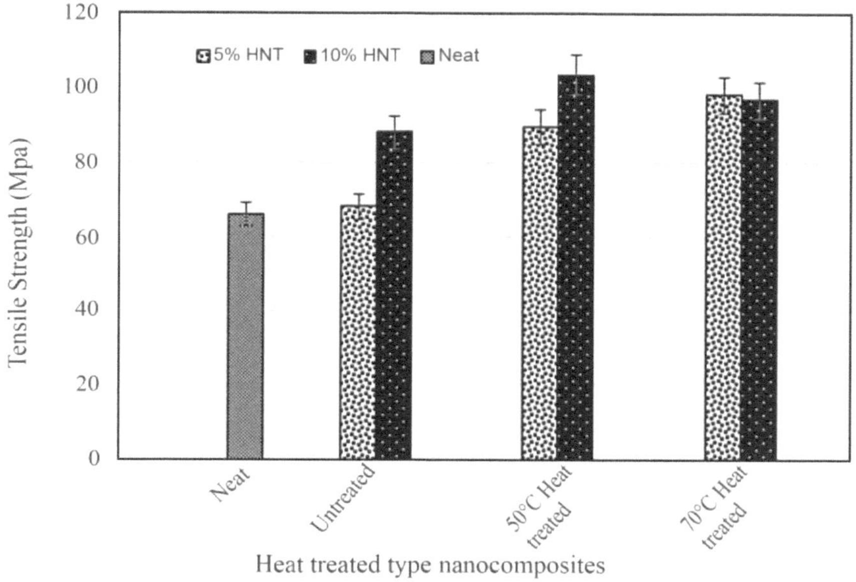

FIGURE 1.7 Tensile strength of epoxy/HNT nanocomposites.

1.3.5 FLEXURAL STRENGTH OF EPOXY/HNT NANOCOMPOSITES

Epoxy/heat-treated HNT reinforced nanocomposites exhibit an increasing trend in flexural strength as shown in Figure 1.8. This is due to a positive effect of HNT on the performance of epoxy resin correlated with the unique characteristics of the HNT to impart better flexural rigidity. The uniform dispersion and a strong interfacial bonding between modified or heat-treated HNT and epoxy register better flexural value compared to neat and untreated composites. There is a drop in flexural strength value of nanocomposites at higher heat-treated temperature (70°C) due to interfacial slippage between nanotubes and epoxy matrix during the bending of the composites due to the inert surface of nanotubes that leads to improvement in the flexural rigidity of the composite specimens at the neutral fiber of the HNTs thereby enhancing its flexural strength.

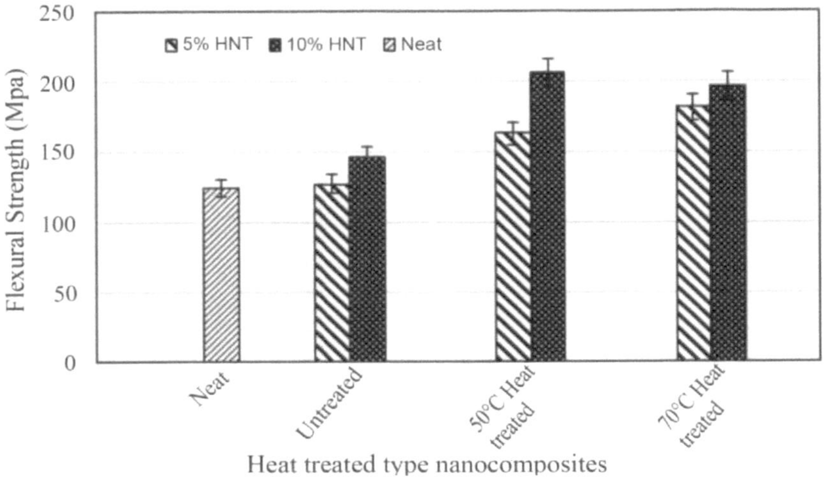

FIGURE 1.8 Flexural strength of epoxy/HNT nanocomposites.

1.3.6 IMPACT ENERGY OF EPOXY/HNT NANOCOMPOSITES

Figure 1.9 shows the variation of the impact energy of epoxy composites with different conditions of HNTs in the composites. Under impact

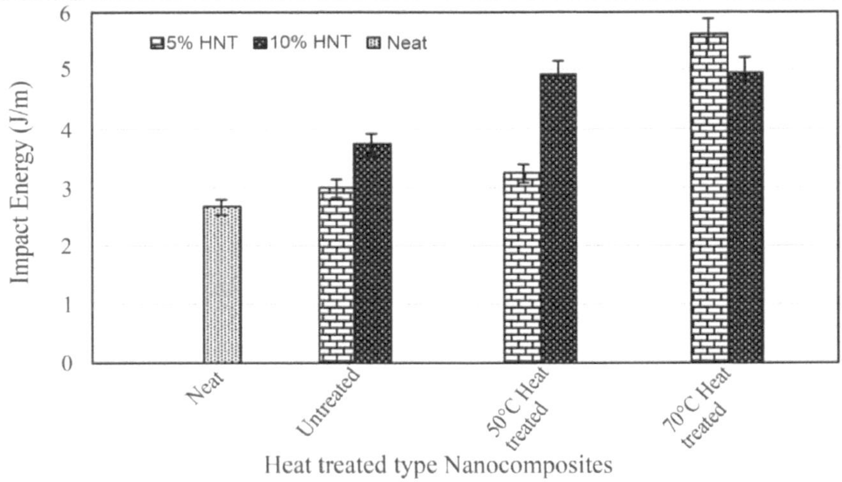

FIGURE 1.9 Impact energy of epoxy/HNT nanocomposites.

loading situation, a gradual improvement in impact strength with the addition of HNT in different Wt.% is noticed in the composites due to better compatibility between epoxy and HNT that will be the important factor to obtain the nanocomposites with a high level of dispersion in order to provide better impact strength and toughness. The impact strength of composite having a 70°C heat treated HNT with 5 Wt.% composition reinforcement in epoxy is higher as compared to other grades of composites. The further higher weight percentage of HNT leads to an increase in brittleness of epoxy resin which decreases the impact strength of the epoxy composites.

1.4 CONCLUSIONS

In this study, the microstructure and mechanical properties are evaluated for heat-treated HNTs synthesized with the epoxy matrix. The primary objective is to generate a strong interfacial bond between the epoxy and the HNTs through an optimal synthetic process. There are three types of heat treatments incorporated for HNTs; viz. untreated HNTs (at RT), 50°C heat-treated HNTs and 70°C heat-treated HNTs. The mechanical stirring process is effectively employed to disperse the HNT in epoxy resin thoroughly. Experimentally, the inferences drawn from the analysis of the results are as follows:

- The epoxy nanocomposites are successfully fabricated reinforcing with HNT particles by polymer stir casting method, the addition of HNT in epoxy nanocomposites exhibits an improvement in the mechanical properties in both untreated and heat-treated conditions.
- The mechanical properties viz. hardness, tensile, and flexural strength are increasing for 50°C heat-treated condition with 10 wt.% of HNT incorporated in the epoxy matrix, Thus it could be proposed as the optimum mixture ratio for epoxy nanocomposites for a number of applications requiring above mentioned attributes.
- An improvement in the impact performance of epoxy nanocomposites is observed with respect to varying filler loading conditions and heat treatment temperatures. The 70°C heat-treated 5 wt.%

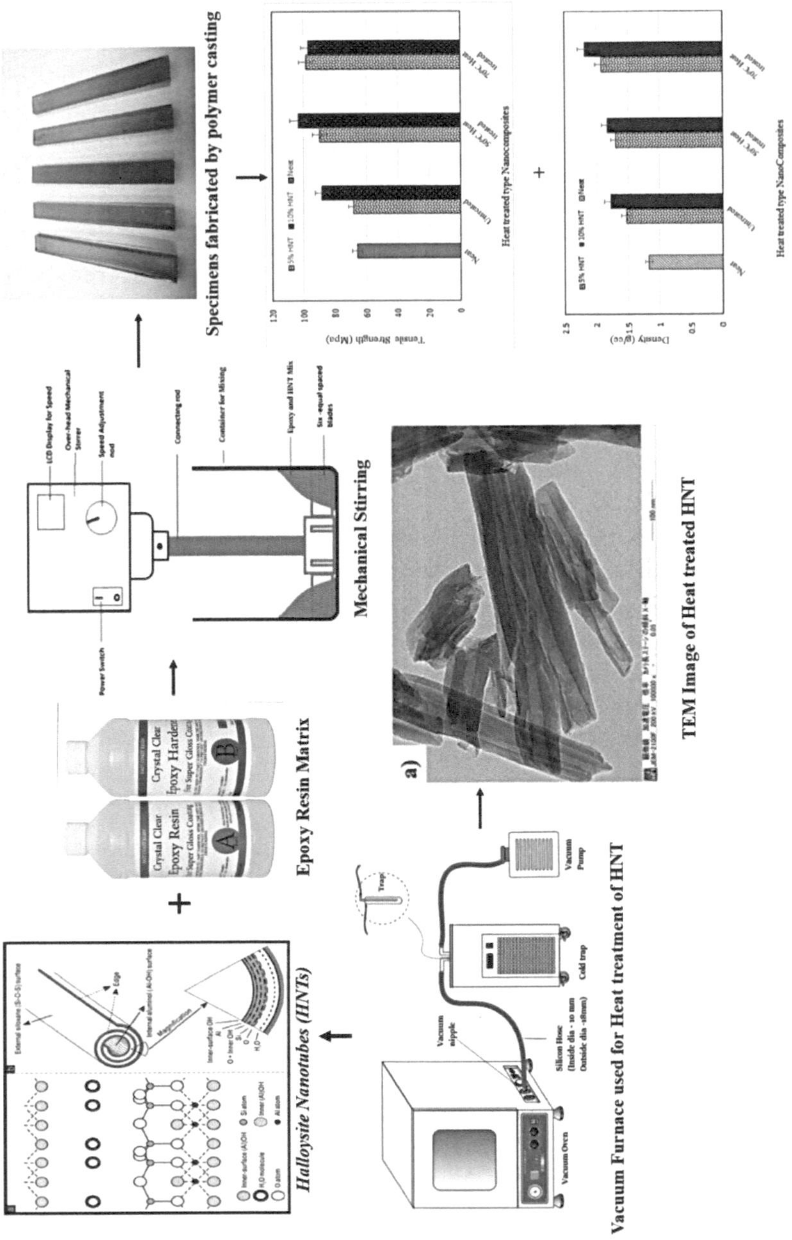

FIGURE 1.10 Graphical illustration.

HNT added to the epoxy nanocomposite registered highest impact resistance value of 5.61 J/m.

- Further, the density and hardness of the composite specimens improved with the addition of reinforcements and incorporation of heat treatment conditions irrespective of loading and other attributes (Figure 1.10).

ACKNOWLEDGMENTS

The authors are grateful to the management of M/s. Brakes India Limited, Nanjangud, and Karnataka, India for providing the required facilities for conducting the present study. The authors are also thankfully acknowledged VTU-RRC Belagavi, Karnataka, India for their co-operation and encouragement to carry out this research work.

KEYWORDS

- **epoxy**
- **halloysite nanotube (HNT)**
- **heat treatment**
- **mechanical properties**
- **polymer stir casting**
- **vacuum furnace**

REFERENCES

1. Joussein, E., Petit, S., Churchman, J., Theng, B., Righi, D., & Delvaux, B., (2005). "Halloysite clay minerals: A review." *Clay Minerals, 40*(4), 383–426.
2. Wang, K., Chen, L., Wu, J. S., Toh, M. L., He, C. B., & Yee, A. F., (2005). "Epoxy nanocomposites with highly exfoliated clay: Mechanical properties and fracture mechanisms." *Macromolecules, 38*(3), 788–800.
3. Wang, J. G., Fang, Z. P., Gu, A. J., Xu, L. H., & Liu, F., (2006). "Effect of amino-functionalization of multiwalled nanotubes on the dispersion with epoxy resin matrix." *J. Applied Polymer Science, 100*(1), 97–104.
4. Deng, S. Q., Ye, L., & Friedrich, K., (2007). "Fracture behaviors of epoxy nanocomposites with nano-silica at low and elevated temperatures." *J. Material Science, 42*(8), 2766–2774.

5. Ye, Y. P., Chen, H. B., Wu, J. S., & Ye, L., (2007). "High impact strength epoxy nanocomposites with natural nanotubes." *Polymer*, *48*(21), 6426–6433.
6. Deng, S. Q., Zhang, J. N., Ye, L., & Wu, J. S., (2008). "Toughening epoxies with halloysite nanotubes." *Polymer*, *49*(23), 5119–5127.
7. Deng, S. Q., Zhang, J. N., & Ye, L., (2009). "Halloysite-epoxy nanocomposites with improved particle dispersion through ball mill homogenization and chemical treatments." *Composites Science and Technology*, *69*(1), 2497–2505.
8. Shi, X. M., Nguyen, T. A., Suo, Z. Y., Liu, Y. J., & Avci, R., (2009). "Effect of nanoparticles on the anticorrosion and mechanical properties of epoxy coating." *Surfaces and Coating Technology*, *204*(3), 237–245.
9. Liu, M. X., Guo, B. C., Du, M. L., Lei, Y. D., & Jia, D. M., (2008). "Natural inorganic nanotubes reinforced epoxy nanocomposites." *J. Polymer Research*, *15*(3), 205–212.
10. Li, C. P., Liu, J. G., Qu, X. Z., Guo, B. C., & Yang, Z. Z., (2008). "Polymer-modified halloysite composite nanotubes." *J. Applied Polymer Science*, *110*(6), 3638–3646.
11. Jia, Z. X., Luo, Y. F., Guo, B. C., Yang, B. T., Du, M. L., & Jia, D. M., (2009). "Reinforcing and flame-retardant effects of halloysite nanotubes on LLDPE." *Polymer-Plastics Technology*, *48*(6), 607–613.
12. Du, M. L., Guo, B. C., & Jia, D. M., (2006). "Thermal stability and flame-retardant effects of halloysite nanotubes on poly (propylene)." *European Polymer Journal*, *42*(6), 1362–1369.
13. Basara, G., Yilmazer, U., & Bayram, G., (2005). "Synthesis and characterization of epoxy-based nanocomposites." *J. Applied Polymer Science*, *98*(3), 1081–1086.
14. Joussein, E., Petit, S., & Delvaux, B., (2007). "Behavior of halloysite clay under formanide treatment." *Applied Clay Science*, *35*(1), 17–24.
15. Yuan, P., Southon, P. D., Liu, Z. W., Green, M. E. R., Hook, J. M., & Antill, S. J., (2008). "Functionalization of halloysite clay nanotubes by grafting with gamma aminopropyl triethoxysilane." *J. Physical Chemistry, C.*, *112*(40), 15742–15751.
16. Liu, M. X., Guo, B. C., Du, M. L., Cai, X. J., & Jia, D. M., (2007). "Properties of halloysite nanotube-epoxy resin hybrids and the interfacial reactions in the systems." *Nanotechnology*, *18*(45), 455703–455709.
17. Kim, G. M., Lee, D. H., Hoffmann, B., Kressler, J., & Stoppelmann, G., (2001). "Influence of nanofillers on the deformation process in layered silicate/polyamide-12 nanocomposites." *Polymer*, *42*(3), 1095–1100.
18. Liu, T. X., Tjiu, W. C., Tong, Y. J., He, C. B., Goh, S. S., & Chung, T. S., (2004). "Morphology and fracture behavior of intercalated epoxy/clay nanocomposites." *J. Applied Polymer Science*, *94*(3), 1236–1244.
19. Brunner, A. J., Necola, A., Rees, M., Gasser, P., Kornmann, X., & Thomann, R., (2006). "The influence of silicate-based nano-filler on the fracture toughness of epoxy resin." *Engineering Fracture Mechanics*, *73*(16), 2336–2345.

CHAPTER 2

STUDY OF PROPERTIES OF NANOSTRUCTURES AND METAL NANOCOMPOSITES ON THEIR BASIS

A. YU. FEDOTOV[1,2] and ALEXANDER V. VAKHRUSHEV[1,2]

[1]*Department of Mechanics of Nanostructures, Institute of Mechanics, Udmurt Federal Research Center, Ural Division, Russian Academy of Sciences, Izhevsk, Russia, E-mail: alezfed@gmail.com*

[2]*Department of Nanotechnology and Microsystems, Technic Kalashnikov Izhevsk State Technical University, Izhevsk, Russia*

ABSTRACT

The mathematical model of condensation processes of metal nanoparticles from a gas phase is submitted. The determination method of mechanical, numerical, and structural nanoparticle properties is proposed.

Theoretical analysis of composites' elastic characteristics including nanoparticles is carried out. Calculation results of metals nanoparticles formation at vacuum evaporation and condensation are given. The numerical and structural properties of forming nanoobjects are examined, and the dependences of structural behavior for nanoobjects and composites are obtained. The research methods considered in the work can be used in relation to nanoparticles and nanocomposites of a broad spectrum of activity, including for predicting the properties of new materials and predicting their behavior during operation.

2.1 INTRODUCTION

Nanostructured and nanoscale elements, structures, and objects have taken a strong place among the new promising materials used in various fields of human activity. These materials attract attention due to their peculiarities and properties at the nanoscale, due to which their performance characteristics and functional behavior distinguish this class of substances from ordinary and customary macro objects in an advantageous way.

The interest of the world industry and science to nanomaterials is confirmed by the growing number of publications, the volume of investments and the number of innovative projects in this field. Nanotechnologies are widely used in mechanical engineering [1] and metalworking [2, 3], the space industry and aircraft industry [4, 5], medicine, and biotechnology [6, 7], the food and textile industry [8, 9], the automotive industry [10, 11], agriculture [12, 13], ecology [14], solar energy and energy-saving [15, 16], and of course, electronics [17, 18].

The study of the processes of interaction and the formation of substances at the nanoscale allows obtaining new functionality materials, as well as keeping track of their properties and structural features to determine their molecular structure. In this situation, the question of the formation of nanoparticles, as well as obtaining homogeneous nano-dispersed mixtures of these and other nanoelements for making nanocomposites with homogeneous and stable in terms of the characteristics of the material is given considerable interest. This is due to the fact that the physical-mechanical, chemical, and other properties of nanoparticles greatly and typically depend nonlinearly on the nanoparticle size.

One of the promising directions for the use of heterophase multicomponent nanosystems is aerosol fire extinguishing generators. Such fire extinguishers are designed to contain and extinguish fires of a wide range of materials; they are universal, generally automated, and have become widespread due to their economic efficiency [19, 20].

The principle of operation of the aerosol gas generator is based on the strong inhibitory effect of nanoaerosol on the combustion reaction of materials in oxygen. Fire extinguishing nanoaerosol generators are effective in the speed of operation, non-toxic, and safe from an environmental point of view. The main advantage of aerosol fire extinguishing nanosystems is the high penetrating power, which is caused by the combined use of gas and fire extinguishing powder. Also not unimportant factors of such gas generators are their dielectric non-conductivity, low cost and chemical neutrality [21]. The appearance of the nanoaerosol gas generator is shown in Figure 2.1.

FIGURE 2.1 Appearance of a nanoaerosol fire extinguishing generator.

The aim is to describe a method of modeling the structural, quantitative, and deformation properties of nanoparticles and composites based on it's throughout the life cycle of nanoparticles starting from forming and finishing phase of use. Value for the practice of the study is that it is related to the calculation of the real process of forming nanoparticles produced according to the modulus of elasticity of the size of the nanoparticles that will allow for the production of micro- and nano-composite materials with the desired customer elastic properties. The results can be used as a basis for further investigations of the elastic properties of nano- and micro-structural materials of any geometric shape.

2.2 MATHEMATICAL MODEL AND THEORETICAL FOUNDATIONS

Isolated nanoparticles are generally prepared by evaporation, thermal saturation and the subsequent condensation of the vapor on or near the cold surface. This synthesis technology is easy to use, cost-effective, and allows nanocrystalline powders on an industrial scale. Devices using the method of evaporation-condensation, a different mode of delivery of the bulk material (powder, solid, liquid), the organization of the heating and cooling of the working environment, the collection of condensed matter. The general principle of operation of such systems is demonstrated in Figure 2.2.

Formation nanoelements may occur in a vacuum or in an inert gas atmosphere, whereby the particles are rapidly losing kinetic energy due to collisions with gas atoms. High-quality nanocomposites using this method are achieved by high-temperature treatment. Cooling the heated composite rate affects the amount of condensation centers, and therefore the formation of nanoparticles and growth rate. An advantage of the evaporation and subsequent condensation is the possibility of using a broad class of nanocomposites.

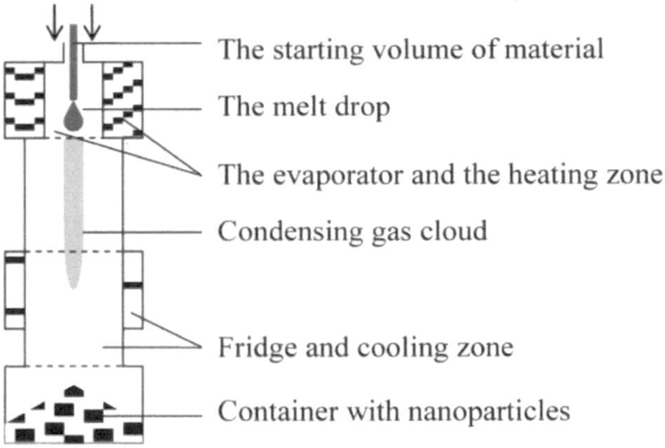

FIGURE 2.2 The scheme of the plant for producing nanoparticles, implementing the principle of thermal saturation and condensation.

Simulation of the formation of nanoparticles synthesis technology that simulates the thermal saturation and subsequent condensation was carried out by molecular dynamics, which is based on the numerical solution of differential equations of motion Newton for each atom with an initial assignment of velocities and coordinates:

$$m_i \frac{d^2 \overline{\mathbf{r}}_i(t)}{dt^2} = -\frac{\partial U(\overline{\mathbf{r}}(t))}{\partial \overline{\mathbf{r}}_i(t)} + \overline{\mathbf{F}}\left(\overline{\mathbf{r}}(t), t\right) \quad (1)$$

$$t = 0, \overline{\mathbf{r}}_i\left(t_0\right) = \overline{\mathbf{r}}_{i0}, \frac{d\overline{\mathbf{r}}_i\left(t_0\right)}{dt} = \overline{\mathbf{V}}_i\left(t_0\right) = \overline{\mathbf{V}}_{i0}, \ i = 1, 2, .., K \quad (2)$$

where K – the number of atoms that formed nanosystem; m_i – the mass of the i-th atom; $\overline{\mathbf{r}}_{i0}, \overline{\mathbf{r}}_i(t)$ – the initial and current radius vector of the i-th atom, respectively; $U(\overline{\mathbf{r}}(t))$ – the potential energy of the system; $\overline{\mathbf{V}}_{i0}, \overline{\mathbf{V}}_i(t)$ – the initial and current speed of the i-th atom, respectively; $\overline{\mathbf{r}}(t) = \{\overline{\mathbf{r}}_1(t), \overline{\mathbf{r}}_2(t), .., \overline{\mathbf{r}}_K(t)\}$ – shows the dependence of the location of all the atoms system, $\overline{\mathbf{F}}(\overline{\mathbf{r}}(t), t)$ – an external force.

Simulation of the formation of nanoaerosol systems was carried out in a representative volume of calculated with periodic boundary conditions. A more detailed statement of the problem and the simulation methodology described in earlier papers [22, 23].

The potential energy of the system (1) is determined by the modified embedded-atom method (MEAM). The theory of MEAM is derived by using density functional theory (DFT). DFT method is currently considered the most recognized approach to the description of the electronic properties of solids. In the embedded-atom method, the complete electron density is a linear superposition of spherically averaged functions. This disadvantage is eliminated in the modified embedded atom method.

The method MEAM energy of the system is written in the form:

$$E = \sum_i \left(F_i \left(\frac{\overline{\rho}_i}{Z_i} \right) + \frac{1}{2} \sum_{j \neq i} \varphi_{ij} \left(R_{ij} \right) \right) \tag{3}$$

where E – atomic energy i; F_i –function for atom i, embedding in the electron density $\overline{\rho}_i$; Z_i – the number of nearest-neighbor atoms i reference in its crystalline structure; φ_{ij} – pair potential between atoms i, j, located at a distance R_{ij}.

In MEAM $F(\rho)$ determined as a function of embedding:

$$F(\rho) = AE_c \rho \ln \rho, \tag{4}$$

where A – controlled variable; E_c – binding energy.

Pair potential between atoms i, j is determined by:

$$\varphi_{ij}(R) = \frac{2}{Z_i} \left\{ E_i^u(R) - F_i \left(\frac{\overline{\rho}_i^0(R)}{Z_i} \right) \right\} \tag{5}$$

where $\overline{\rho}_i^0$ – electron density.

The total electron density at the dive includes angular dependence and written in the form:

$$\overline{\rho} = \rho^{(0)} G(\Gamma) \tag{6}$$

There are many types of functions $G(\Gamma)$ [16]:

$$G(\Gamma) = \sqrt{1+\Gamma}, \quad G(\Gamma) = e^{\frac{\Gamma}{2}}, \tag{7}$$

$$G(\Gamma) = \frac{2}{1+e^{-\Gamma}}, \quad G(\Gamma) = \pm\sqrt{|1+\Gamma|}. \tag{8}$$

The function Γ is calculated according to the formula:

$$\Gamma = \sum_{h=1}^{3} t^{(h)} \left(\frac{\rho^{(h)}}{\rho^{(0)}} \right)^2 \tag{9}$$

where $h = 0 - 3$, correspond to *s, p, d, f* symmetry; $t^{(h)}$ – weight multipliers; $\rho^{(h)}$ – values, determining the deviation of the distribution of the electron density distribution of a perfect crystal of the cubic system $\rho^{(0)}$:

$$s(h=0): \rho^{(0)} = \sum_i \rho^{a(0)}(r^i), \tag{10}$$

$$p(h=1): \left(\rho^{(1)}\right)^2 = \sum_{\alpha} \left[\sum_i \rho^{a(1)}(r^i) \frac{r_{\alpha}^i}{r^i} \right]^2, \tag{11}$$

$$d(h=2): \left(\rho^{(2)}\right)^2 = \sum_{\alpha,\beta} \left[\sum_i \rho^{a(2)}(r^i) \frac{r_{\alpha}^i r_{\beta}^i}{r^{2i}} \right]^2 - \frac{1}{3} \sum_i \left[\sum_i \rho^{a(2)}(r^i) \right]^2, \tag{12}$$

$$f(h=3): \left(\rho^{(3)}\right)^2 = \sum_{\alpha,\beta,\gamma} \left[\sum_i \rho^{a(3)}(r^i) \frac{r_{\alpha}^i r_{\beta}^i r_{\gamma}^i}{r^{3i}} \right]^2. \tag{13}$$

Here $\rho^{a(h)}$ – radial functions that represent a decrease in the contribution of distances r_i, superscript i indicates the nearest atoms, α,β,γ – summing the codes for each of the three possible directions. Finally, the individual contribution is calculated according to the formula:

$$\rho^{a(h)}(r) = \rho_0 e^{-\beta^{(h)}\left(\frac{r}{r_e}-1\right)} \tag{14}$$

The dynamic observation of the elastic properties of the nanocomposites was implemented as follows. Matrix composites with nano inclusions were subjected to shear or tensile strain (Figure 2.3). Here, as the input parameters of the system set the speed of deformation. After the effect on the nanosystem, its atoms begin to rebuild. The main mechanical and physical properties were calculated.

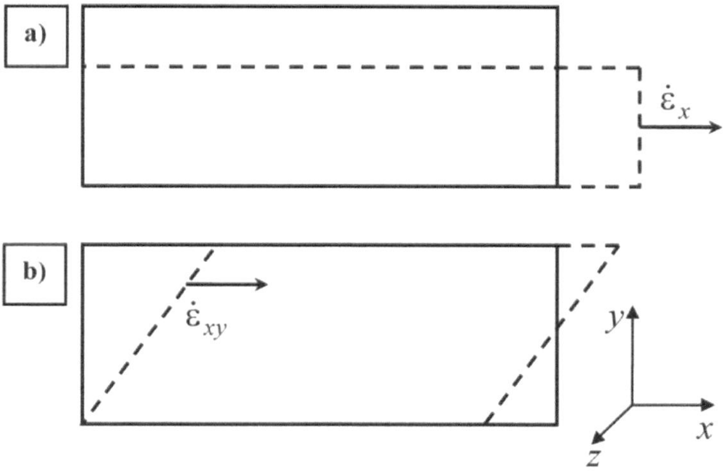

FIGURE 2.3 Scheme nanocomposite stretching deformation (a) and shear (b).

The main deformation characteristics of objects include effective bulk modulus of the material k and the effective shear modulus μ. If necessary, the transition is possible from these values to Young's modulus E and Poisson's ratio ν according to Eq. (16). Backward calculations are possible too.

$$E = \frac{9k\mu}{3k+\mu}, \nu = \frac{3k-2\mu}{2(3k+\mu)}. \tag{15}$$

When working with composites and studying its properties, bulk material generally been considered the two components: the matrix and inclusions. The matrix is a macro material forming the primary composite structure. The inclusions are some additives in the matrix to refine and improve its mechanical or other characteristics. Deformation properties of the matrix are compatible with the bulk material. They are usually taken from reference books. The strength properties of the inclusions are much more difficult to define. Either experimental data or theoretical methods are used when calculating them. In our case, as the inclusions are nanoparticles. To determine the deformation properties and dependencies of the model of the elastic "equivalent" element tensile concentrated forces described above are used.

It is known from previously conducted studies [17] and the effect of the Hall-Petch, the effective bulk modulus and effective shear modulus of nano-objects is not a linear function of their size. Generalized dependence of these quantities can be written in the form of expressions:

$$k_I = \begin{cases} A\left(\frac{d}{d_0}\right)^B, \text{if } d \le d_0 \\ k_{I0}, \text{if } d > d_0 \end{cases} = \begin{cases} A\left(\bar{d}\right)^B, \text{if } \bar{d} \le 1 \\ k_{I0}, \text{if } \bar{d} > 1 \end{cases} \tag{16}$$

$$\mu_I = \begin{cases} M\left(\frac{d}{d_0}\right)^L, \text{if } d \le d_0 \\ \mu_{I0}, \text{if } d > d_0 \end{cases} = \begin{cases} M\left(\bar{d}\right)^L, \text{if } \bar{d} \le 1 \\ \mu_{I0}, \text{if } \bar{d} > 1 \end{cases} \tag{17}$$

where k_I, μ_I – the effective bulk modulus and effective shear modulus of inclusions, $d, \bar{d}$ – actual and relative sizes of the nanoparticles, k_{I0}, μ_{I0} – the effective bulk modulus and effective shear modulus of volume material, which consists of inclusions, d_0 – the diameter of the nanoobjects in which they cease to depend on the elastic properties of the size, *A*, *B*, *M*, *L* – some of the coefficients, calculated empirically. Having determined the values of the coefficients *A*, *B*, *M*, *L*, it is possible to calculate the strength parameters of nanostructures of all sizes.

In some cases, it is convenient to use dimensionless or relative deformation characteristics of the inclusions. Dimensionless made in relation to macro properties of the material k_{I0} and μ_{I0}:

$$\bar{k}_I = \frac{k_I}{k_{I0}} = \begin{cases} \frac{A}{k_{I0}} \left(\bar{d}\right)^B, \text{if } \bar{d} \le 1 \\ 1, \text{if } \bar{d} > 1 \end{cases} \tag{18}$$

$$\bar{\mu}_I = \frac{\mu_I}{\mu_{I0}} = \begin{cases} \frac{M}{\mu_{I0}} \left(\bar{d}\right)^L, \text{if } \bar{d} \le 1 \\ 1, \text{if } \bar{d} > 1 \end{cases} \tag{19}$$

With known mechanical parameters and dependencies for inclusions in the form of nanoparticles, the calculation of similar properties of the composite material performed based on formulas and techniques from [18]. Model of the medium with a low volume fraction of spherical inclusions is expressed by the following relationships:

$$\bar{k} = \frac{k}{k_M} = 1 + \frac{\left(\frac{k_I}{k_M} - 1\right)c}{1 + \left(k_I - k_M\right) \Big/ \left(k_M + \frac{4}{3}\mu_M\right)} \tag{20}$$

$$\bar{\mu} = \frac{\mu}{\mu_M} = 1 - \frac{15\left(1 - \nu_M\right)\left[1 - \left(\mu_I / \mu_M\right)\right]c}{7 - 5\nu_M + 2\left(4 - 5\nu_M\right)\left(\mu_I / \mu_M\right)} \tag{21}$$

where $k, \bar{k}$ – the dimension and the relative effective bulk modulus of the composite, $\mu, \bar{\mu}$ – the dimension and the relative effective shear modulus of the composite, c – the volume fraction of the nanoparticles included in the composite material, k_M, μ_M – the deformation parameters of the matrix.

2.3 HARDWARE BASE AND DESCRIPTION OF DEVICES

Experimental studies of nanostructures were primarily aimed at the comparison with theoretical data obtained using mathematical modeling. Therefore, the main objects of study were the dimensional, surface, and spatial properties of clusters formed in nanocomposites.

As a hardware base, we used the complex system for measuring the properties of materials Nanotest 600, shown in Figure 2.4, and the NewView 6300 contactless optical profilometer, shown in Figure 2.5. This equipment allows a comprehensive multi-parameter analysis of nanostructures of aerosol composition.

FIGURE 2.4 System for measuring and analyzing the properties of nanomaterials NanoTest 600.

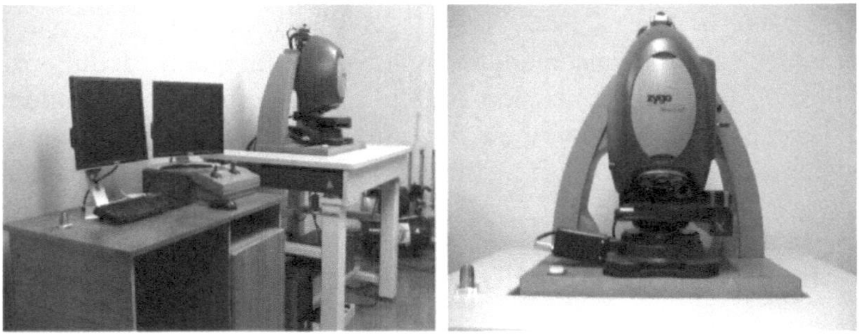

FIGURE 2.5 Image of (a) contactless optical profilometer NewView 6300 and (b) of its base module.

The complex system for measuring the properties of materials Nanotest 600 is designed to study the properties of materials, including physic mechanical, at the nano- and microscale. This system is able to work in static and dynamic modes and allows you to get an image of the studied sample, determine the elastic modulus, hardness, friction coefficient, surface roughness, work expended on elastic-plastic deformation, as well as other mechanical characteristics. Nanotest 600 has a resolution of 0.04 nm in depth, the resolution in the horizontal plane is 20 nm.

The non-contact optical profilometer New View 6300 is used to obtain a three-dimensional picture and the surface relief of materials at the nanoscale. The principle of operation of the profilometer refers to light microscopy. In order to obtain an image of the sample under study and measure its spatial topology, the NewView 6300 scans the sample in a non-contact manner based on the interference of white light. The light inside the interferometric lens is divided into two beams: one beam is reflected from the sample under study, while the other is reflected from the control surface located inside the lens. In the future, both light beams are directed into a semiconductor chamber, which converts light signals into electrical ones.

The imposition of two light waves leads to the appearance of light and dark interference fringes on the sample. By means of strips, the structure of the sample surface is analyzed. Scanning of the entire sample area is achieved by vertical movement of the lens by a piezoelectric motor. The resulting light pattern is processed in the camera and converted into a digital image consisting of dots of different brightness. Vertical measurements are performed according to interference data. Measurements in the plane of the scan sample are based on the field of view and resolution of the lens used. The results of the surface profile of the sample are displayed on the display in the form of images, graphs, and numerical data. The profilometer has a vertical resolution of up to 0.1 nm, the horizontal one varies in the range of 0.43–11.6 μm and depends on the selected lens.

2.4 RESEARCH RESULTS AND ANALYSIS

Formation of composite metal nanoparticles was investigated for the one-, two-, and three-component nanoparticles. In all cases, the condensation step in the metal nanoclusters atoms proceeded actively. In the process of grouping involved all types of initial materials. The structure of the nanoobjects was obtained solid. Cavities were not observed in nanoparticles. Nanostructures are obtained predominantly spherical.

Consider the formation of nanoparticles of an example wherein starting metals studied ternary mixtures of silver, gold, and zinc. The mass fraction of each metal in the nanosystem was selected approximately equal to the following values: Ag – 33.97%, Zn – 37.05%, Au – 28.98%. Phase condensation of metal atoms in nanoparticles, following the heat, was simulated for 30 ns. The grouping of the atoms in the nanoclusters is actively carried out in the first moments of time and was accompanied by the formation of a significant amount of nanoparticles. Later on, the condensation already formed nanoobjects is observed, which leads to a gradual decrease in the number of nanoparticles and an increase in their size (Figure 2.6).

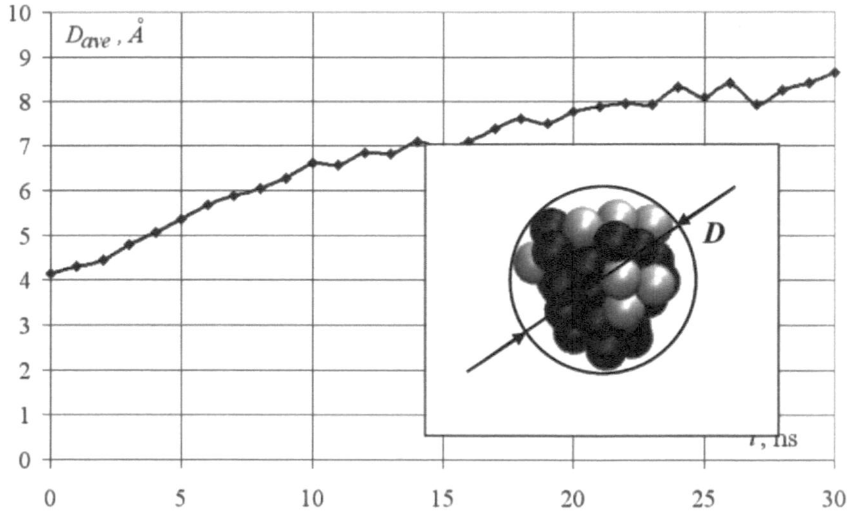

FIGURE 2.6 Changing of the average diameter of the nanoparticles formed from the 3-component metal mixture in a calculating cell during the condensation phase.

Analysis of the internal structure of layers of nanoparticles was carried out on samples having typical characteristics for all nanoelements. We determined the particle radius and diameter, and then detected the structure and composition of each layer of the nanoparticles as a function of the relative radius of the nanostructure. Graph of the relative density of the layers nanoparticles is shown in Figure 2.7. The total value of the relative density of each layer was assumed to be 100%. Internal analysis nanoelements showed an uneven distribution of metal nanoparticles in the structure under study. The core of the particle consists mainly of gold, the middle layers are formed by atoms of silver, and zinc atoms form a shell. There are transition layers in which there are several metals.

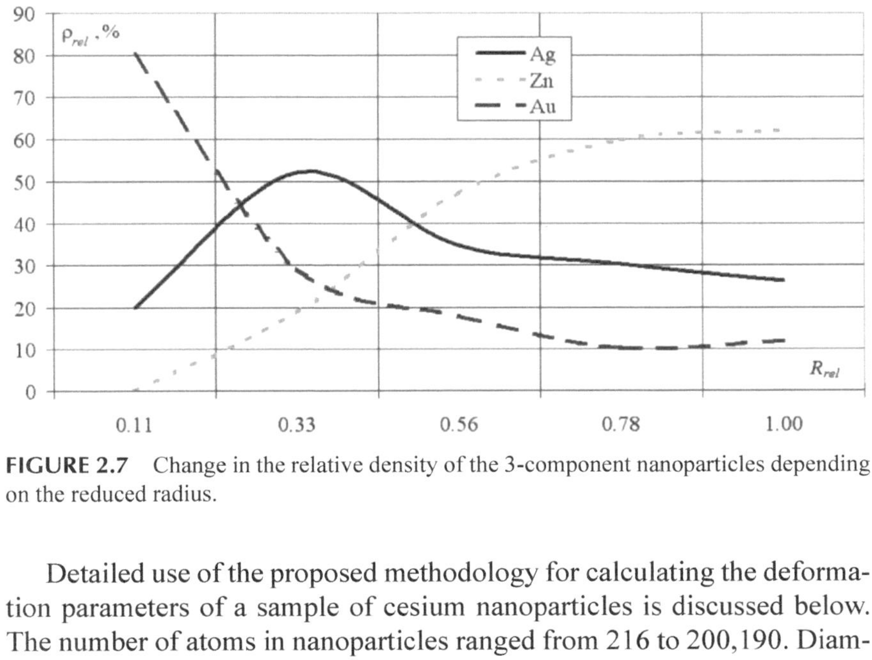

FIGURE 2.7 Change in the relative density of the 3-component nanoparticles depending on the reduced radius.

Detailed use of the proposed methodology for calculating the deformation parameters of a sample of cesium nanoparticles is discussed below. The number of atoms in nanoparticles ranged from 216 to 200,190. Diameter equilibrium nanoparticles with cesium are from 3 nm to 27 nm. The number of atoms is increased as long as the elastic modulus reaches a value nanoparticle reference value Cs. An important task in the development of nanotechnology is to determine the size of the nanoparticles, in which the mechanical characteristics will be the same as the reference values of the macro stuff. Of particular interest to the study of mechanical characteristics of the nanoparticles are "small" size nanostructures with the number of atoms of <10,000. The mean square error is a pronounced minimum (Figure 2.8). We calculate the bulk modulus of nanoparticles of cesium from the system [7].

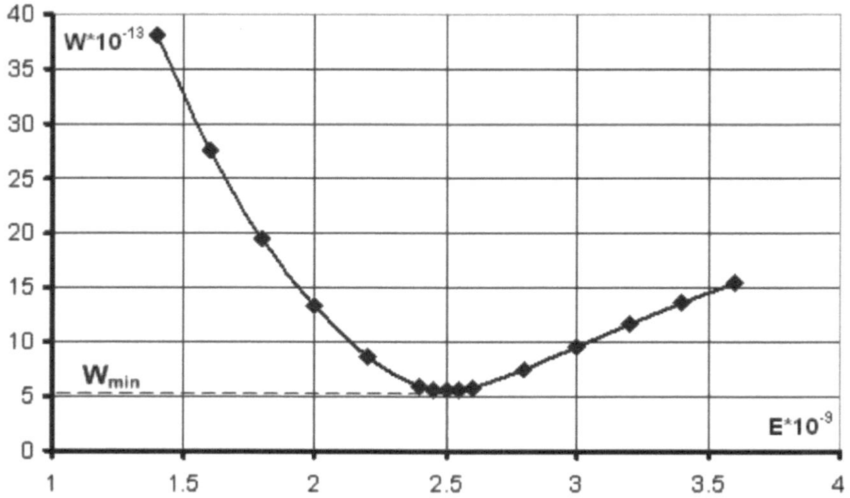

FIGURE 2.8 The dependence of the mean squared error W of the elastic Young's modulus E "equivalent" element, for nanoparticles of cesium (49 995 atoms).

Changing the effective bulk modulus dimensionless inclusions depending on the diameter of the above nanoparticles according to Eq. (19) is shown in Figure 2.9. Triangular markers are shown the data series of numerical experiments, the results of which were subsequently constructed approximating the trend line. The mathematical expression of the curve approximation allowed us to determine unknown coefficients $A = 0.882$ and $B = -1.083$ of the Eq. (2.19). The index of the effective bulk modulus indicates the magnitude of Poisson's ratio corresponding to the reference value of cesium.

The mathematical formulation of the curve of the dimensionless bulk modulus of inclusions is the basis for the calculation of the deformation characteristics of composite materials containing nanoparticles. Bulk modulus and shear modulus of the composite depends on the size of the inclusions contained therein. For a composite material consisting of a polystyrene matrix and nanoinclusions cesium dependence of the dimensionless unit of the reduced diameter is shown in Figure 2.10.

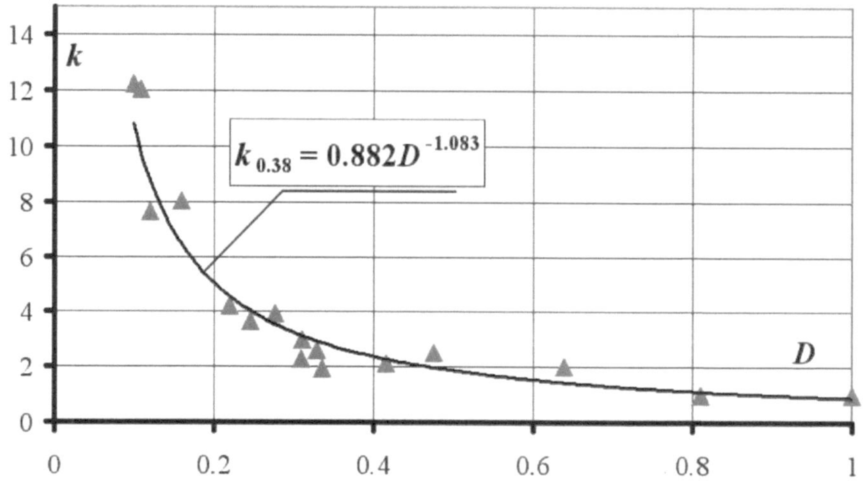

FIGURE 2.9 Changing the dimensionless effective bulk modulus inclusions depending on the dimensionless diameter of inclusions.

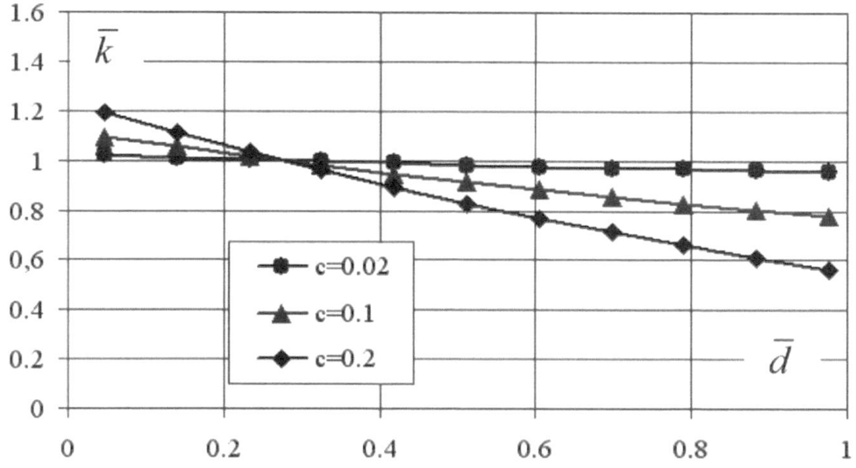

FIGURE 2.10 The dependence of the dimensionless bulk modulus of the nanocomposite of the reduced size of the nanoparticles contained in it.

Analysis of the chart shows that the deformation characteristics are largely dependent on the proportion of the concentration of nanoparticles in the formed composite. Small nanostructures $(\bar{d} < 0,3)$ increase the bulk modulus of the nanocomposite, large nanoparticles ($\bar{d} > 0,3$) lead to its weakening.

Further studies were carried out for the following materials: pure aluminum; aluminum nanocomposite with iron nanoparticles radius of 1 nm; aluminum nanocomposite nanocylinders iron radius of 1 nm. Periodic boundary conditions were used for the simulation. The simulation was performed under standard thermodynamic conditions.

The process of the tensile strain of the nanocomposites under study and of a sample of pure aluminum is shown in Figures 2.11–2.19. Tension patterns are presented for various modeling points.

Figures 2.11–2.13 illustrate the stretching of a pure aluminum volume element. For the initial times shown in Figure 2.11, it is characterized by the active restructuring of the atomic structure of the sample. Atoms are displaced relative to the sites of the crystal lattice. Their coordinate positions are asymmetrical. The forces arising within the sample at this stage are not balanced.

The subsequent behavior of the pure aluminum sample for the simulation time of 17 psec and 25 psec shows that structural changes at the atomic level continue. New energetically favorable states appear. Also, some ordering for the coordinate positions of atoms begins to be seen. Observed sample stretching. The cross-section decreases with its elongation. Full rupture and destruction at this stage does not occur.

A sample of an aluminum matrix with an iron nanoparticle under tension behaves differently. The stretching patterns of this sample for the same simulation time points are presented in Figures 2.14–2.16.

The initial rearrangement of the positions of atoms in Figure 2.14 noticeable for aluminum. The iron atoms forming the nanoparticle change their coordinates slightly, due to the stronger interaction potential between them.

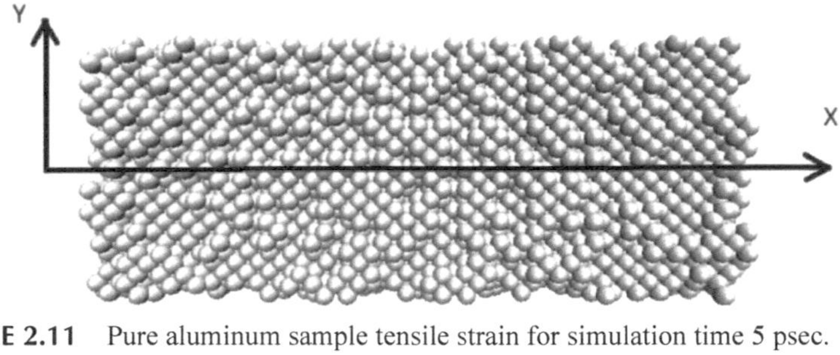

FIGURE 2.11 Pure aluminum sample tensile strain for simulation time 5 psec.

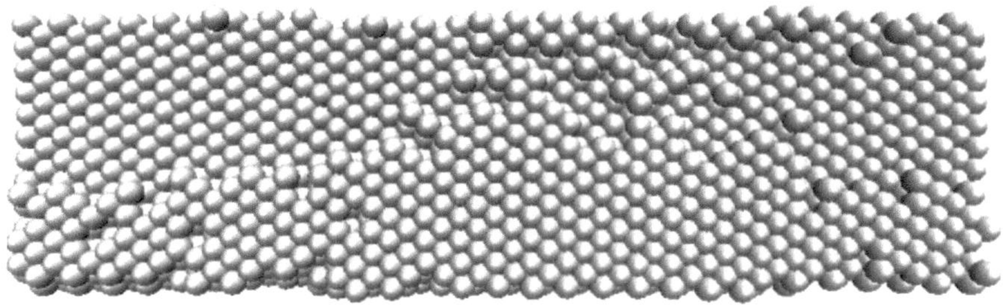

FIGURE 2.12 Pure aluminum sample tensile strain for simulation time 17 psec.

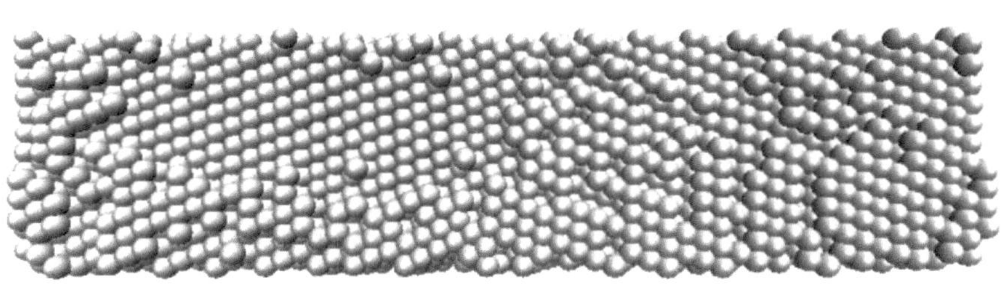

FIGURE 2.13 Pure aluminum sample tensile strain for simulation time 25 psec.

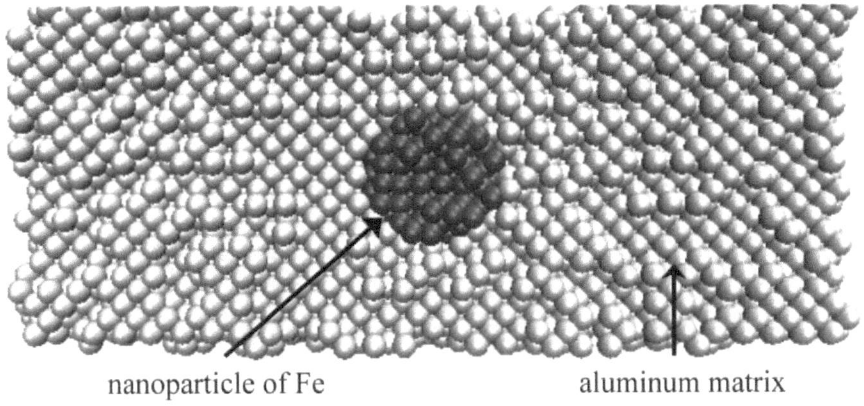

FIGURE 2.14 Tension deformation of an aluminum nanocomposite with an iron nanoparticle for a simulation time of 5 psec.

The destruction of an aluminum nanocomposite with an iron nanoparticle occurs directly in the region of the nanoparticle. This is evident from Figures 2.15 and 2.16. There is a detachment of the aluminum matrix from the surface of the nanoparticle, followed by rupture of the sample. Until the nanocomposite is completely destroyed, a thin bridge of aluminum atoms is noticeable.

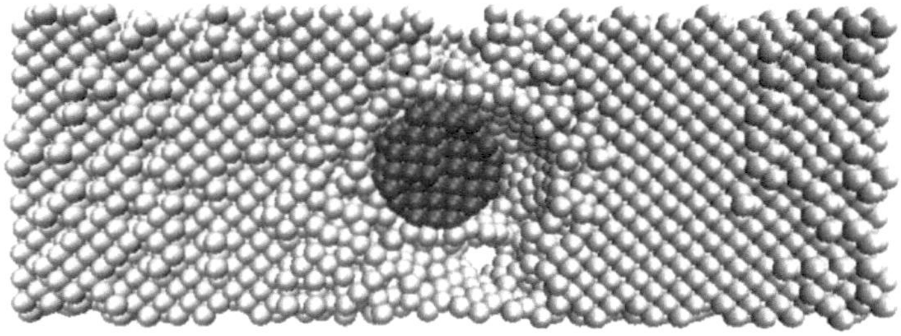

FIGURE 2.15 Tension deformation of an aluminum nanocomposite with an iron nanoparticle for a simulation time of 17 psec.

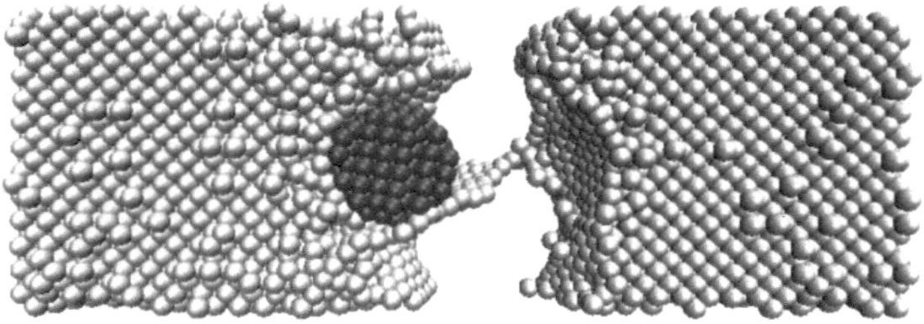

FIGURE 2.16 Tension deformation of an aluminum nanocomposite with an iron nanoparticle for a simulation time of 25 psec.

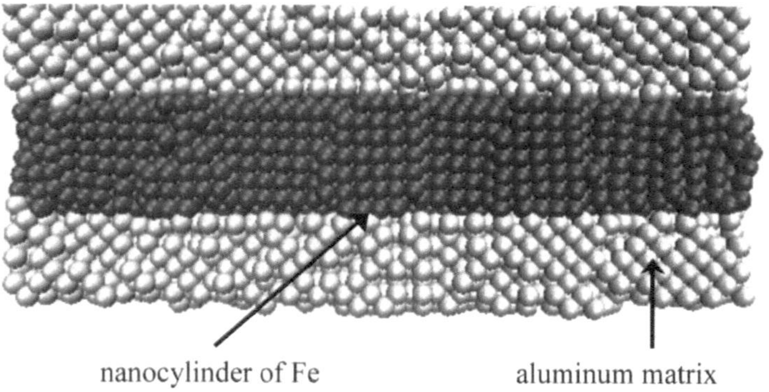

FIGURE 2.17 Tension deformation of an aluminum nanocomposite with an iron nanocylinder for a simulation time of 5 psec.

The stretching of the sample from the matrix of aluminum with an iron nanorod for different simulation times is shown in Figures 2.17–2.19. As for the case of a sample of pure aluminum for a nanocomposite with a rod of complete destruction does not occur. However, for the time of 25 psec, there is a distortion and a strong restructuring of the coordinate positions of aluminum atoms in the central zone. In the lateral areas, the violation of the crystal structure is less noticeable. For all three samples, mid stress and mid strain plots were plotted in the direction of tension (Figure 2.20).

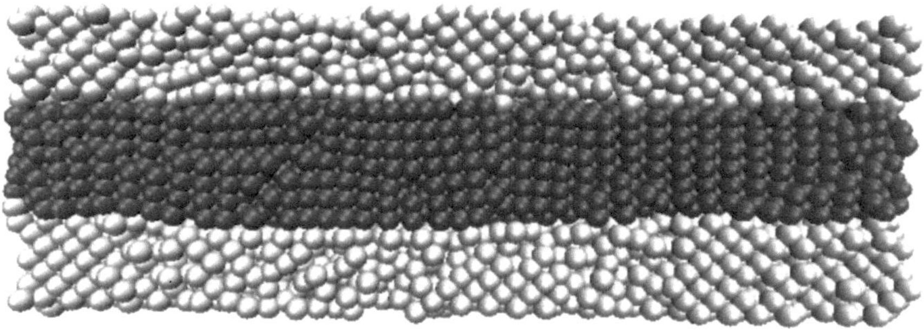

FIGURE 2.18 Tension deformation of an aluminum nanocomposite with an iron nanocylinder for a simulation time of 17 psec.

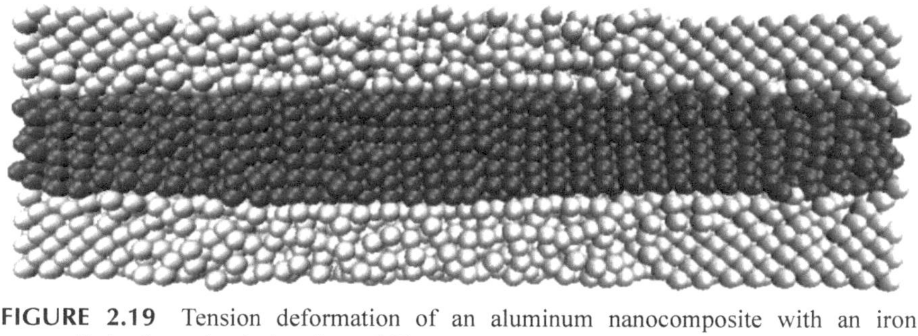

FIGURE 2.19 Tension deformation of an aluminum nanocomposite with an iron nanocylinder for a simulation time of 25 psec.

The process of shear deformation in the *x* and *y* plane of the studied nanocomposites and the sample of pure aluminum in Figures 2.21–2.29 is shown. Shift patterns for various modeling times are presented. Figures 2.21–2.23 illustrate the shear deformation of a pure aluminum volume element. In comparison with the tensile strain, the asymmetric arrangement of aluminum atoms is observed for all the simulation time points. Atoms are displaced relative to the sites of the crystal lattice.

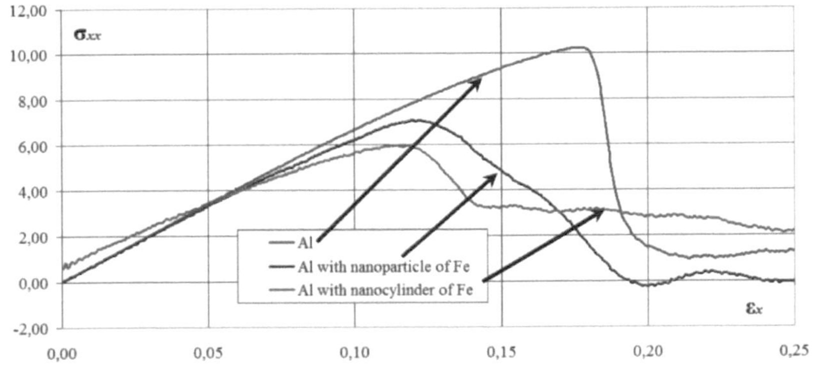

FIGURE 2.20 The mid-stress-strain curves for a variety of nanocomposite materials.

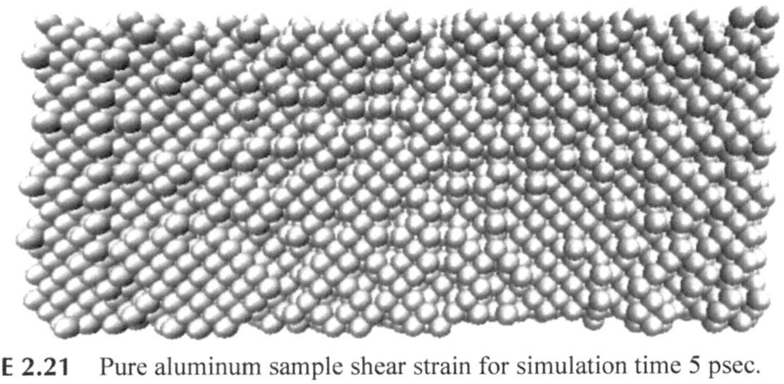

FIGURE 2.21 Pure aluminum sample shear strain for simulation time 5 psec.

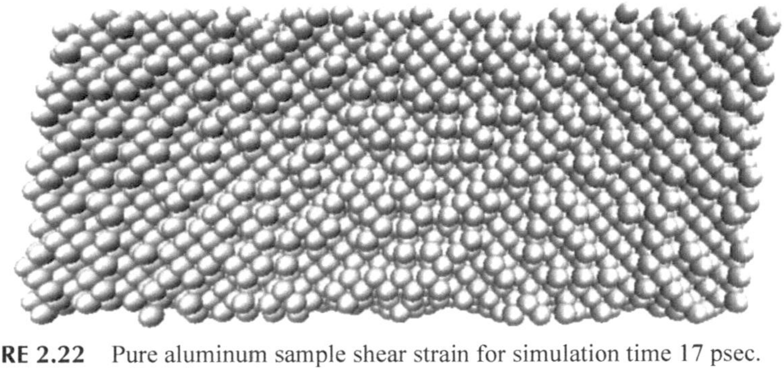

FIGURE 2.22 Pure aluminum sample shear strain for simulation time 17 psec.

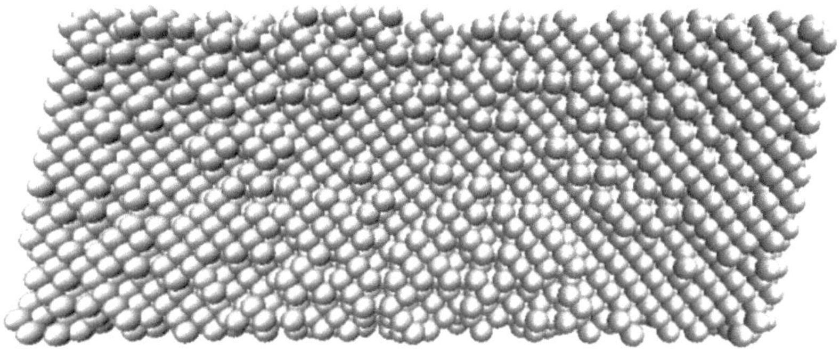

FIGURE 2.23 Pure aluminum sample shear strain for simulation time 25 psec.

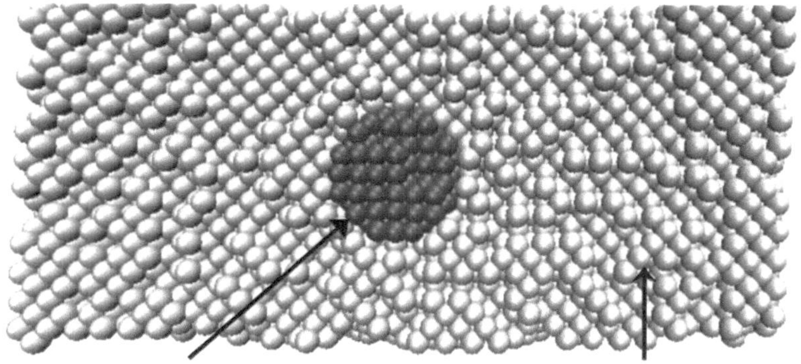

FIGURE 2.24 Shear deformation of an aluminum nanocomposite with an iron nanoparticle for a simulation time of 5 psec.

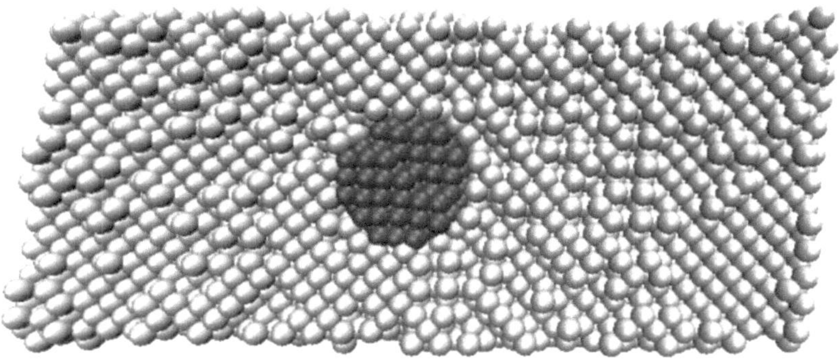

FIGURE 2.25 Shear deformation of an aluminum nanocomposite with an iron nanoparticle for a simulation time of 17 psec.

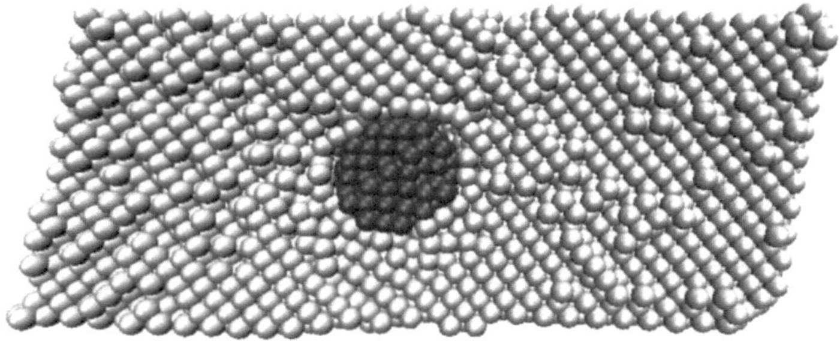

FIGURE 2.26 Shear deformation of an aluminum nanocomposite with an iron nanoparticle for a simulation time of 25 psec.

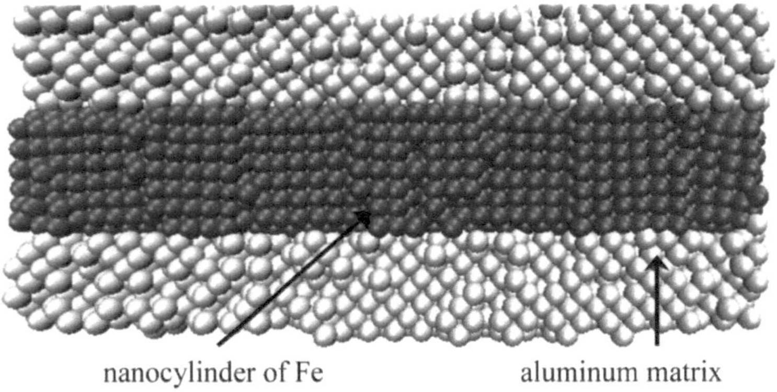

FIGURE 2.27 Shear deformation of an aluminum nanocomposite with an iron nanocylinder for a simulation time of 5 psec.

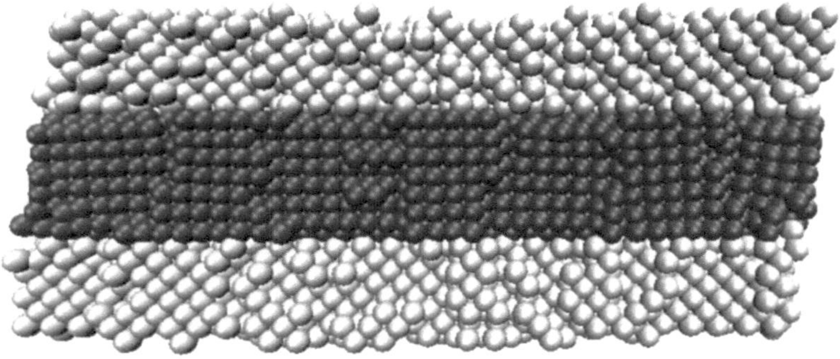

FIGURE 2.28 Shear deformation of an aluminum nanocomposite with an iron nanocylinder for a simulation time of 17 psec.

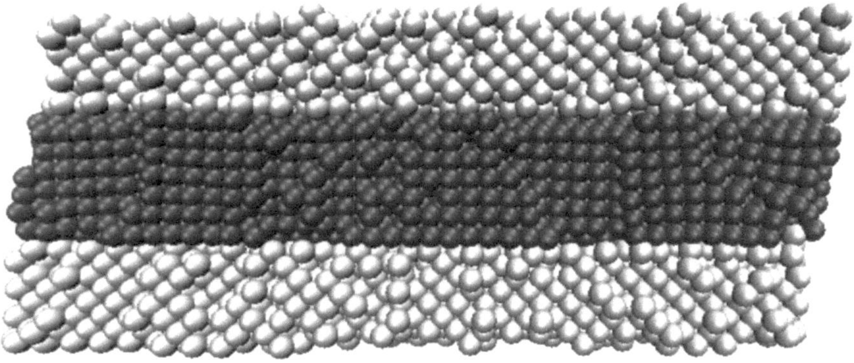

FIGURE 2.29 Shear deformation of an aluminum nanocomposite with an iron nanocylinder for a simulation time of 25 psec.

The shear strain patterns for samples with a nanoparticle and an iron nanocylinder for the same simulation times are shown in Figures 2.24–2.29. In these cases, aluminum atoms are more actively rebuilt. Iron atoms that belong to nano-inclusions have a layered displacement. For shear deformation for all samples, their complete destruction was not observed. For the studied samples, we plotted the mid stress and mid strain curves in the shear direction (Figure 2.30).

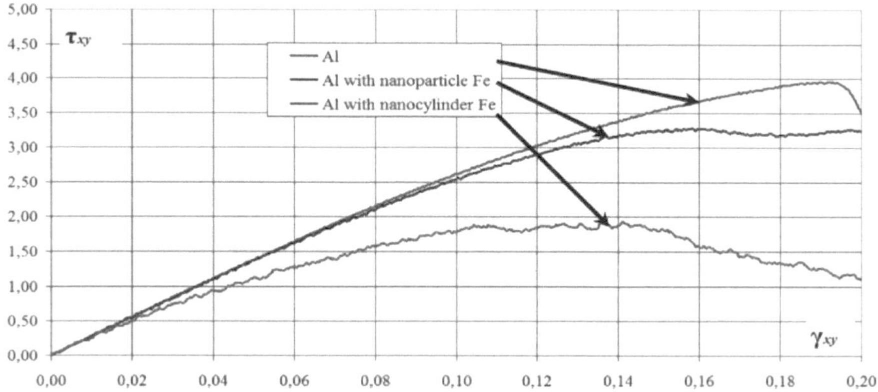

FIGURE 2.30 The mid-stress-strain curves for shearing a variety of nanocomposite materials.

2.5 CONCLUSION

The simulation method of structural, quantitative, and deformation properties of nanoparticles and composites based on them is offered. The model allows studying the dynamic characteristics of nanoobjects throughout the life cycle of their use, starting from the processes of the formation of nanoparticles and ending its influence on the mechanical parameters of the composite.

Calculated dependences of the modulus of elasticity, bulk modulus of nanoparticles of cesium on their diameter were obtained. Curves were obtained on the basis of matching the solutions of the molecular dynamics and the theory of elasticity, carried by vectors of displacements in points coinciding with the position of the atoms of the nanoparticle. The critical diameter of nanoparticle cesium (21.5 nm) is calculated. If you increase, the size of nanostructures is more critical diameter the strength characteristics of nanoparticles coincide with the reference values of the macro materials.

The hardware base and devices for studying the properties of nanoparticles and nanocomposites based on them are described. The principles of operation of the complex measuring system NanoTest 600 and contactless optical profilometer NewView 6300 are given.

The dependence of the dimensionless effective bulk modulus of nanoparticle cesium is built. Based on this the mathematical formulation of the dependence of the effective bulk modulus is given. Formalization of the change in the effective bulk modulus and shear modulus nanoinclusions depending on their size is an important part to determine the deformation properties of nanocomposite materials.

On the example of a polystyrene matrix and cesium nanoparticles, it is shown that adding nanostructures in the composite can lead to different effects. The hardening of the material is observed at small sizes of nanostructures. The deterioration of the strength characteristics occurs when the nanoparticles have a significant size.

ACKNOWLEDGMENTS

The works were carried out with financial support from the Research Program of the Ural Branch of the Russian Academy of Sciences (project 18-10-1-29).

KEYWORDS

- **elastic properties**
- **math modeling**
- **mechanical properties**
- **modified embedded-atom method**
- **molecular dynamics**
- **nanocomposites**
- **nanoparticles**

REFERENCES

1. Regan, B. C., Aloni, S., Jensen, K., Ritchie, R. O., & Zettl, A., (2005). Nanocrystal-powered nanomotor. *Nano Letters*, *5*(9), 730–1733.
2. Dutta, A. K., Narasaiah, N., Chattopadhyay, A. B., & Ray, K. K., (2002). Influence of microstructure on wear resistance parameter of ceramic cutting tools. *Materials and Manufacturing Processes*, *17*(5), 651–670.
3. Gaitonde, V. N., Karnik, S. R., Figueira, L., & Davim, J. P., (2009). Analysis of machinability during hard turning of cold work tool steel. *Materials and Manufacturing Processes, 23*(4), 1373–1382.
4. Barnett, D. M., Rawal, S. P., & Rummel, K., (2001). Multifunctional structures for advanced spacecraft. *Journal of Spacecraft and Rockets*, *38*(2), 226–230.
5. Okpala, C. C., (2014). The benefits and applications of nanocomposites. *International Journal of Advanced Engineering Technology*, *5*(4), 12–18.
6. Emerich, D. F., & Thanos, C. G., (2003). Nanotechnology and medicine. *Expert Opinion on Biological Therapy*, *3*(4), 655–663.
7. Darroudi, M., Ahmad, M. B., Abdullah, A. H., Ibrahim, N. A., & Shameli, K., (2010). Effect of accelerator in green synthesis of silver nanoparticles. *Int. J. Mol. Sci.*, *11*(10), 3898–3905.
8. Sawhney, A. P. S., Condon, B., Singh, K. V., Pang, S. S., Li, G., & Hui, D., (2008). Modern applications of nanotechnology in textiles. *Textile Research Journal*, *78*(8), 731–739.
9. Wong, Y. W. H., Yuen, C. W. M., Leung, M. Y. S., Ku, S. K. A., & Lam, H. L. I., (2006). Selected applications of nanotechnology in textiles. *Autex Research Journal*, *6*(1), 1–10.
10. Louda, P., (2007). Applications of thin coatings in automotive industry. *Journal of Achievements in Materials and Manufacturing Engineering*, *24*(1), 51–56.
11. Wong, K. V., & Paddon, P. A., (2014). Nanotechnology impact on the automotive industry. *Recent Patents on Nanotechnology*, *8*(3), 181–199.

12. Gindl, W., Schöberl, T., & Jeronimidis, G., (2004). The interphase in phenol-formaldehyde and polymeric methylene diphenyl-di-isocyanate glue lines in wood. *Int. J. Adhes. Adhes.*, *24*(4), 279–286.
13. Bhatnagar, A., & Sain, M., (2005). Processing of cellulose nanofiber-reinforced composites. *Journal of Reinforced Plastics and Composites*, *24*(12), 1259–1268.
14. Zakharov, R. S., & Glotov, O. G., (2007). The burning characteristics of pyrotechnic compositions with powdered titanium. *Bulletin of the NSU. Series: Physics*, *2*(3), 32–40.
15. Ma, H. Y., Bendix, P. M., & Oddershede, L. B., (2012). Large-Scale orientation dependent heating from a single irradiated gold nanorod. *Nano Lett.*, *12*(8), 3954–3960.
16. Sanchot, A., Baffou, G., Marty, R., Arbouet, A., Quidant, R., Girard, C., & Dujardin, E., (2012). Plasmonic nanoparticle networks for light and heat concentration. *ACS Nano*, *6*(4), 3434–3440.
17. Boyer, D., Tamarat, P., Maali, A., Lounis, B., & Orrit M., (2002). Photo-thermal imaging of nanometer-sized metal particles among scatterers. *Science*, *297*(5584), 1160–1163.
18. Shipway, A. N., Katz, E., & Willner, I., (2000). Nanoparticle arrays on surfaces for electronic, optical, and sensor applications. *Chem. Phys. Chem.*, *1*(1), 18–52.
19. Baratov, A. N., Korolchenko, A. Y., Kravchuk, G. N., et al., (1990). *Fire and Explosion Hazard of Substances and Materials and Means of Their Suppression* (p. 496). Moscow: Chemistry, (in Russian).
20. Sinilov, V. G., (2004). *Security, Fire and Fire Alarm Systems* (2nd edn., p. 352), Moscow: Publishing Center "Academy" (in Russian).
21. Terebnev, V. V., Artemyev, N. S., & Korolchenko, D. A., (2006). *Fire Protection and Extinguishing Fires* (p. 412). Industrial buildings and structures, Moscow: Pozhnauka, (in Russian).
22. Vakhrushev, A. V., Fedotov, A. Y., Vakhrushev, A. A., Golubchikov, V. B., & Givotkov, A. V., (2011). Multilevel simulation of the processes of nano aerosol formation. Part 1: Theory foundations. *International Journal of Nano Mechanics Science and Technology*, *2*(2), 105–132.
23. Vakhrushev, A. V., Fedotov, A. Y., Vakhrushev, A. A., Golubchikov, V. B., & Givotkov, A. V., (2011). Multilevel simulation of the processes of nanoaerosol formation. Part 2: Numerical investigation of the processes of nanoaerosol formation for suppression of fires. *International Journal of Nanomechanics Science and Technology*, *2*(3), 205–216.

CHAPTER 3

POSSIBLE MECHANISM OF REDOX SYNTHESIS OF METAL/CARBON NANOCOMPOSITES MODIFIED BY P, D ELEMENTS

V. I. KODOLOV,[1,2] V. V. TRINEEVA,[1,3] I. N. SHABANOVA,[1,3] and N. S. TEREBOVA[1,3]

[1]*Basic Research – High Educational Center of Chemical Physics and Mesoscopics, Studencheskaya str. 7, 423067, Izhevsk, Russia, E-mail: vkodol.av@mail.ru*

[2]*M.T. Kalashnikov Izhevsk State Technical University, Studencheskaya str. 7, 423067, Izhevsk, Russia, E-mail: kodol@istu.ru*

[3]*Udmurt Federal Research Center, Ural Division, RAS, T. Baramzina str. 34, 423066, Izhevsk, Russia, E-mail: vera_kodolova@mail.ru*

ABSTRACT

The redox synthesis results of modified metal/carbon nanocomposites in which metal may be Copper or Nickel, and the applied modifiers—ammonium polyphosphate, silicon, sodium thiosulfate, metal oxides (Metal: Al, Fe(3), Cu(2) and Ni) were analyzed at the relation changes of modifier and nanocomposite. The hypothesis of redox synthesis with the use of chemical mesoscopic idea about the charge quantilization is proposed. The hypothesis is that the negative charge quants move to reducing element nucleolus to provoke the positive charge quants and the creation of annihilation phenomenon. Therefore, the direct electromagnetic field formed promotes to d electrons shift on higher energetic levels including the carbon shell of nanocomposite nanogranyl. The picture of this phenomenon may be presented:

1. At the joint grinding of modifier with metal/carbon nanocomposites nanogranul the nanoreactor (multiplet) is formed on the boundaries of which the chemical potentials difference creates.
2. Because of the potentials difference, the delocalized electrons, which are disposed on the nanogranul shell, move to positively charged atoms of modifier and stimulate the positive charges quantiliization.
3. The annihilation phenomenon arises at the superposition of quants of negative and positive charges on the boundary of reagents. In this case, the electron magnetic field which activates the delocalization of d electrons of nanogranul metals and the electron shift to nanogranul carbon shell.
4. The electron shift process within nanogranul is facilitated by modified agent polarization growth.

The hypothesis explains the growth of metal atomic magnetic moments which may be bigger than 4.5 μ_B because of the process of delocalization d electrons. The shift of electrons on nanogranul carbon shells leads to the restoration of the electron balance after redox synthesis. There is a correlation between electron number participated in redox synthesis and the value of atomic magnetic moment as well as between the polarization degree of reactive systems and the increasing of atomic magnetic moments of nanogranul metals.

3.1 INTRODUCTION

Earlier the mechanochemical methods of metal/carbon nanocomposites redox synthesis as well as the mechanochemical modification by such p, d elements as aluminum, silicium, phosphorus, sulfur, iron, and nickel, copper are considered [1–5]. It's shown that at the modification the redox process leads to the growth of magnetic moment of metal within nanogranul. This result may be explained by separating of d electrons and then the shift of them on more high energetic levels. For the explanation of this phenomenon, the following hypothesis may be proposed. It's known that during the redox processes the positive and negative charges are appeared. According to the Chemical Mesoscopics, the quantization of charges arises. When quants of positive charges set to quants of negative charges, the annihilation phenomenon takes place. This paper is dedicated to the consideration of facts which may be confirming the hypothesis proposed.

3.2 REASONS CAUSED MECHANISM OF MODIFICATION PROCESS

3.2.1 THE ANALYSIS OF COPPER AND NICKEL/CARBON NANOCOMPOSITES (NI/CNC) STRUCTURE

The metal/carbon nanocomposites represent, according to [2], metal-containing clusters with increased magnetic moments of metals which are located within the carbon shell. The carbon shell consists of carbon fibers included carbine and polyethylene fragments with delocalized electrons at the juncture of fragments connections.

This notion about metal/carbon nanocomposites follows from results analysis of its investigations of roentgenograms, x-ray photoelectron spectra, and also from EPR, TEM, and AFM researches. Correspondent pictures set below (Figures 3.1–3.4)

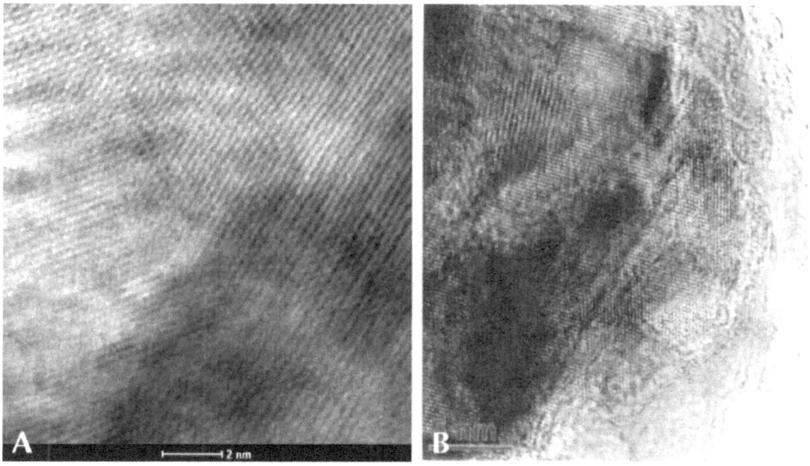

FIGURE 3.1 TEM microphotograph of structures for Cu/C (a) and Ni/C (b) nanocomposites.

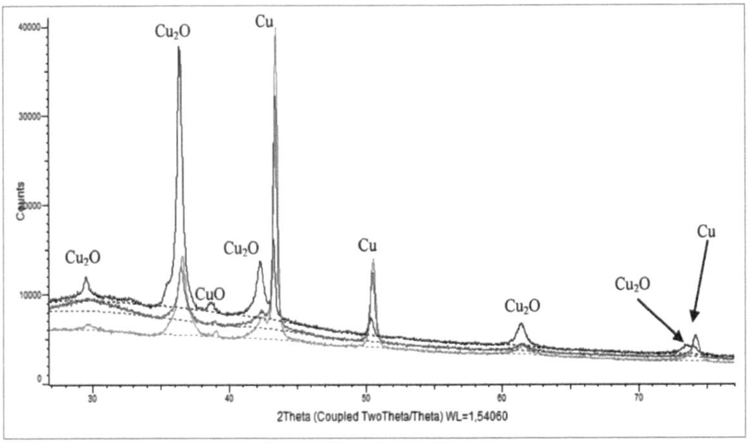

FIGURE 3.2 Roentgenogram of copper/carbon nanocomposite with different content of copper.

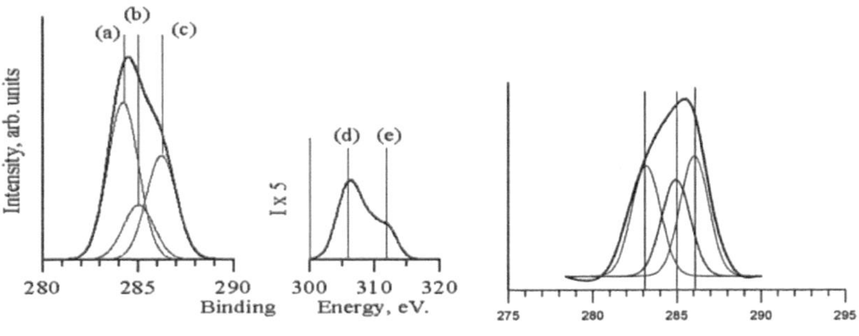

FIGURE 3.3 C1s spectra for copper (a) and nickel/carbon (b) nanocomposites.

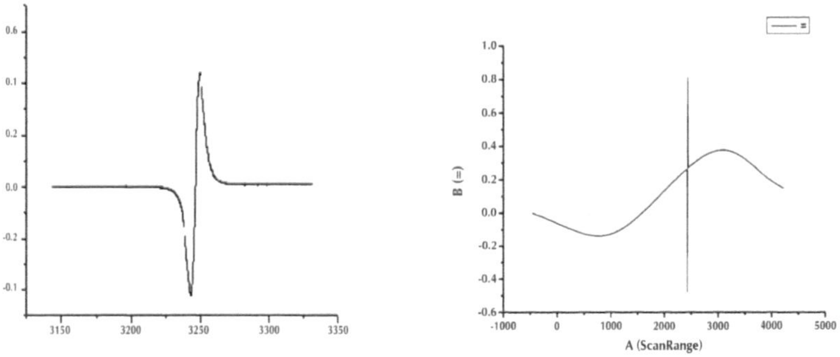

FIGURE 3.4 EPR spectra of nanogranuls of copper (a) and nickel/carbon nanocomposites.

On the base of Van Fleck's theory, the model of calculation is proposed and the metal magnetic moments are calculated.

It's necessary to note that, according to EPR spectra, the spin quantity on carbon shell for copper/carbon nanocomposite (Cu/CNC) equals to 10^{17} spin/gram, and for Ni/CNC – 10^{24} spin/gram.

3.2.2 SHORT COMMUNICATIONS ABOUT MODIFICATION PROCESSES

The modification process is realized on phases boundary in the presence of a small quantity of active component (for example, water) by means of a mechanochemical method that is to say by joint grinding (in mechanical pounder) of mixtures containing nanogranuls of metal/carbon nanocomposites and correspondent substances. Such substances as aluminum oxide, silicon, ammonium polyphosphate, iron (3) oxide, nickel oxide, and copper (2) oxide are used for investigations. The relation of metal/carbon nanocomposites and modifiers is changed from 1:1 to 1:0.2 (nanocomposite: modifier). As a result of the mechanochemical process, the xerogel is formed. The product obtained is heated to 75°C and then to 150°C in vacuum. The powder obtained is activated for sample preparation which is used for correspondent investigations.

3.3 ANALYSIS OF MODIFIED NANOCOMPOSITES INVESTIGATION RESULTS

The presence of active carbon fibers with delocalized electrons in carbon shell of metal/carbon nanocomposites points the way to the interaction nanogranul with modifiers containing positive charged atoms. The quants of electron waves are formed because of the difference of chemical potentials between nanocomposites nanogranuls and the active surface of modifiers (nanoparticles). These quants of negative charges activate the quantization of positive charges of modifier agent atoms. At the superposition of negatively charged quants on positively charged quants, the annihilation phenomenon appears together with the accompanied direct magnetic field. In this case, the magnetic field formed acts to d orbital of nanogranul cluster metal that leads to d electron ripping (delocalization).

This process is accompanied by the reduction of modifier positive charged atoms. It's necessary to say that the reduction of positively charged atoms of modifiers takes place in all cases of modification investigated. These facts are confirmed by the researches results with the application of x-ray photoelectron spectroscopy and roentgenograms. These investigations are added by the transition electron microscopy (TEM) as well as by the electron paramagnetic resonance (EPR). For instance, with the use of TEM of high resolution, it's determined that the shell from carbon fibers for nanogranul of Cu/CNC modified by Ammonium Polyphosphate is practically conserved (Figure 3.5). However, the increased thickness has appeared on the carbon fibers. This is explained by the phosphorus-containing group's implantation within the shell that leads to an increase of nanogranul volume. On the base of EDS, spectra of secondary roentgen radiation activation in the surface layer of nanogranul the possible phases with Phosphorus reduced are determined. These data are conformed to the results of roentgenogram analysis.

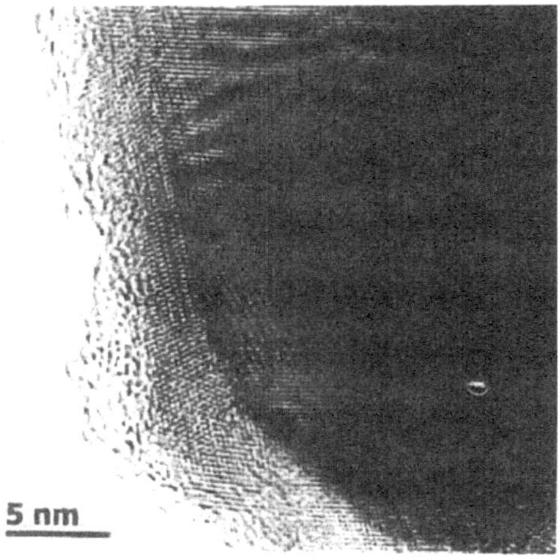

FIGURE 3.5 TEM microphotograph of phosphorus-containing copper/carbon nanocomposite.

According to roentgenogram data, there is the interatomic distance correspondent to Cu–P (θ=43°). Ammonium Phosphate is related to Copper Carbon nanocomposite as 0,5:1, 1:1, 1,5:1 that is corresponded the following peak relations: 1,0; 0,47; 0,17.

The intensity decreasing is caused by the masking of the interaction of phosphorus-containing groups and copper clusters. It's necessary to note the growth of halo in small angles (2–12°) on roentgenograms of phosphorus-containing Cu/C nanocomposites (relation of APPh: Cu/CNC, equaled to 0.5:1). It's interesting that the magnetic moments for Copper at the relations 1:0.5 and 1:1 differ in two times (Table 3.1).

TABLE 3.1 Parameters of Multiplet Splitting of Cu3s Spectra for Cu/C Nanocomposites Modified by Ammonium Popyphosphate, Including Copper Atomic Magnetic Moments

Sample	I_2/I_1	Δ, eV	μ_{met}, μ_B
$Cu_3s_{нано}$	0.2	3.5	1.3
$Cu_3s_{нано}$ (P)	0.4	3.5	2.0
$Cu_3s_{нано}$ (P*2)	0.4	3.5	2.0
$Cu_3s_{нано}$ (P*1/2)	0.85	3.5	4.2

This fact is possibly correlated with regulated phases formation because of the polarization of the phosphorus-containing layer.

The increase of copper atomic magnetic moment is confirmed by the growth of the quantity of spin/gram from 10^{17} to 10^{19} (data of EPR spectra).

P2p spectra for Cu/C nanocomposites modified by different quantities of Ammonium Polyphosphate are presented below (Figure 3.6).

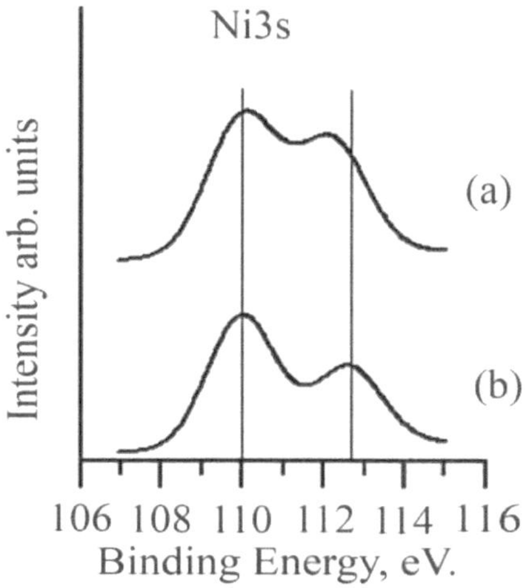

FIGURE 3.6 P2p spectra: a) Cu/C + APPh 1:1; b) Cu/C + APPh 1:0.5.

According to the P2p spectra of modified Cu/C nanocomposites the reduction process at the relation Cu/C + APPh 1:0.5 proceeds on 50% completely in comparison with the process at the relation Cu/C + APPh 1:1. In this case, the bonding energy for Phosphorus corresponds to 130 eV, and the bonding energy equaled to 132,3 eV may be related to P=C bond [6].

Analysis of C1s spectra of modified nanocomposite at the relation Cu/C + APPh 1:0,5 (Figure 3.7) shows the possibility of carbon shell deformation with the increasing of polyacetylene fragments in the surface layer.

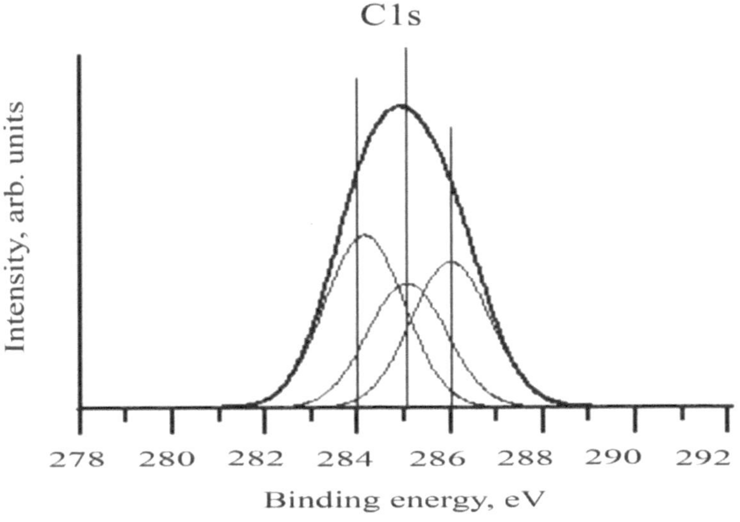

FIGURE 3.7 C1s spectrum for phosphorus-containing nanocomposite obtained at the relation Cu/C + APPh 1:0,5.

This is possible therefore in the comparison with the C1s spectrum of Cu/CNC (Figure 3.3a) the line intensity for CH groups in modified nanocomposite increases three times. Besides the intensities values of sp^2 and sp^3 constituents of spectrum are near that may indicate on spherical form of nanogranul.

In another case, the change of relation of nanocomposite nanogranul/modifier agent at the Cu/CNC modification by Silicon has not influenced on values of Copper atomic magnetic moment (Table 3.2)

TABLE 3.2 Parameters of Multiplet Splitting of Cu3s Spectra for Cu/C Nanocomposites Modified by Silicon, Including Copper Atomic Magnetic Moments

Sample	I_2/I_1	Δ, eV	μ_{met}, $\mu_{Б}$
$Cu3s_{нано}$ (Si)	0.6	3.0	3.0
$Cu3s_{нано}$ (Si 1/2)	0.6	3.0	3.0

According to Si2p spectra ($E_{(SiO)}$ = 101eV; $E_{(Si)}$ = 99eV) reduction process proceeds on 50% and does not depend on the relation changes (Figure 3.8). This is probably explained by the decrease of electron quantization in the silicon layer.

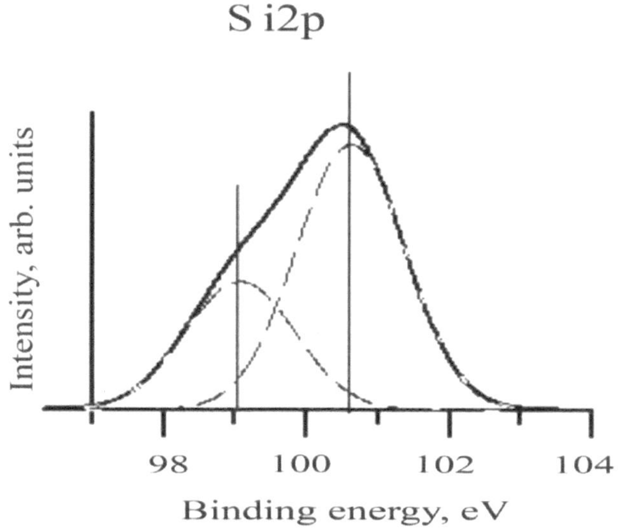

FIGURE 3.8 Si2p spectrum for the interaction product of Cu/CNC + SiO_2 1:1.

If the metal oxide is applied as a modifier, it's possible the competition of metal reduction processes between modifier metal and nanocomposite nanogranul metal. For instance, at the Cu/CNC modification by Nickel oxide, it's observed the atomic magnetic moments growth for Copper (cluster metal within nanogranul) on 0,7 μ_B, and for Nickel – 0,5 μ_B (Table 3.3).

TABLE 3.3 Parameters of Multiplet Splitting of Ni3s and Cu3s Spectra for Cu/C Nanocomposites Modified by Nickel Oxide, Including Atomic Magnetic Moments of Metals

Sample	I_2/I_1	Δ, eV	μ_{met}, μ_B
$Cu3s_{nano}$ (Ni)	$0{,}4_{Cu}/0{,}4_{Ni}$	$3_{Cu}/2_{Ni}$	$2_{Cu}/2{,}3_{Ni}$

Besides the deformation of carbon, the shell takes place as well as the sp hybridization appearance (carbine fragments) (Figure 3.9).

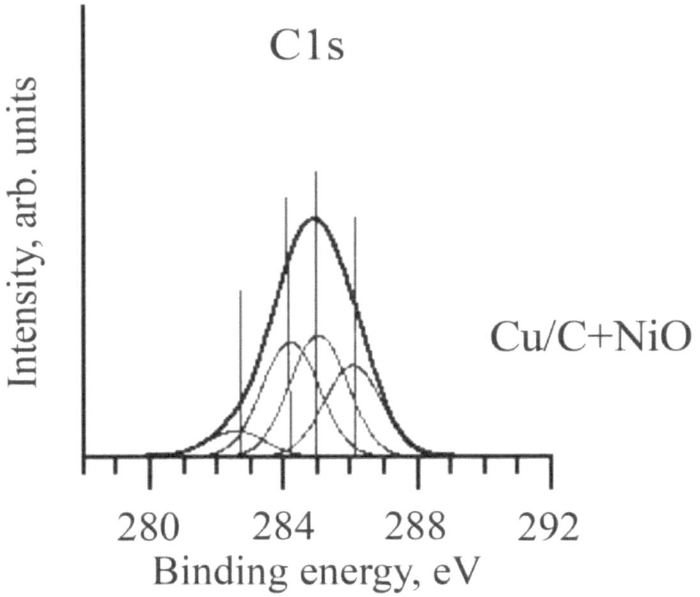

FIGURE 3.9 C1s spectrum of nickel-containing of copper/carbon nanocomposite.

The proposed hypothesis supposes the differences at the modifications of copper/carbon and Ni/CNC because of a higher quantity of delocalized electrons on the carbon shell of Ni/CNC: Ni/CNC – 10^{24} spin/g in comparison with Cu/CNC – 10^{17} spin/g. However, the investigation of the Ni/CNCmodification shows an identical character of metal magnetic properties changes. In Table 3.4, the atomic magnetic moments for Ni/C nanocomposites modified by Si, P, and S containing substances in comparison with modified Cu/C nanocomposites are demonstrated.

TABLE 3.4 The Metal Atomic Magnetic Moment Values Within Nanogranuls Modified by Copper/Carbon and Nickel/Carbon Nanocomposites (the Relation of Reagents at Modification 1:1)

Modified Nanocomposite	$\mu_{Cu,}$	$\mu_M-\mu_0$	Modified Nanocomposite	μ_{Ni}	$\mu_M-\mu_0$
Cu/C–Si	3.0	1.7	Ni/C–Si	4.0	2.2
Cu/C–P	2.0	0.7	Ni/C–P	3.0	1.2
Cu/C–S	1.8	0.5	Ni/C–S	2.8	1.0

μ_M– metal atomic magnetic moment for modified nanocomposite; μ_0 – metal atomic magnetic moment for no modified nanocomposite.

The values of metal atomic magnetic moments may be stipulated by the polarization value of bonds that participated in the redox process. In this case, the polarization of bonds is decreased from Si–O across P–O to S–O bond. It's necessary to note that the influence of the correspondent modifier on the metal atomic magnetic moment is constant: the difference of $\mu_m-\mu_0$ within row (Si, P, S) is not changed, and the difference of metal atomic magnetic moments between series is equaled to 1 μ_B.

The metal atomic magnetic moment changes at Nickel/Carbon modification by the following metal oxides CuO, NiO, Fe_2O_3, and $Al_2O_3$are adduced in Table 3.5.

TABLE 3.5 Atomic Magnetic Moments of Metals in Modified Nickel/Carbon Nanocomposite

Nanocomposite and Modified Agent (Mass Part)	μ_{Ni}, μ_B	μ_{Cu}, μ_B	$\mu_{Fe,}$ μ_B
Ni/C–NiO (0.5)	3,0		
Ni/C–NiO	4,5		
Ni/C–CuO	2,0	2,0	
Ni/C–Fe_2O_3	2,5		3,0
Ni/C–Al_2O_3 (0.2)	4,8		

On the base of the results of Ni/C nanocomposite modification, it's possible to conclude that the ability to the polarization of metal oxide layer influences on the Nickel atomic magnetic moment growth. For example, the significant growth of the Nickel magnetic moment takes place if the Aluminum Oxide is used as a modifier. During this process, the Aluminum is completely reduced (data of x-ray photoelectron spectroscopy).

3.4 CONCLUSION

In the submitted manuscript, the hypothesis of metal/carbon nanocomposites interaction with modifiers on the solid phase's boundaries is proposed. In these conditions of processes the growth of metal atomic magnetic moments because of d electron delocalization, which is explained by the annihilation of positive and negative charges quants, takes place. At the beginning of processes during the joint grinding of components, the multiples (nanoreactors) are formed and simultaneously the flows of quants of negative and positive charges are appeared. In this case, there is a phenomenon of annihilation with the formation of a direct electromagnetic field which stimulates the atomic magnetic moment growth.

KEYWORDS

- **annihilation**
- **charge quantilization**
- **mechanism**
- **mechanochemical grinding**
- **modified nanocomposites**
- **modifiers**
- **positive and negative charge quants**
- **redox synthesis**

REFERENCES

1. Ivashkin, Y. A., (2008). *Defects of Crystal Structure of Deformed Cubic Crystals* (p. 137). Bryansk: BGITL.
2. Kodolov, V. I., & Trineeva, V. V., (2017). New Scientific Trend – Chemical Mesoscopics. *Chemical Physics and Mesoscopy*, *19*(3), 454–465.
3. Shabanova, I. N., Kodolov, V. I., Terebova, N. S., & Trineeva, V. V., (2012). *X-ray Photoelectron Spectroscopy for Investigation of Metal/Carbon Nanosystems and Nanostructured Materials* (p. 252). M.-Izhevsk: Publ. "Udmurt University."
4. Lipanov, A. M., Kodolov, V. I., Melnikov, M. Y., Trineeva, V. V., & Pergushin, V. I., (2016). Influence of minute quantities of metal/carbon nanocomposites on the polymeric materials properties. *Doklady of RAS, 466*(1), 15–17.

5. Kodolov, V. I., Trineeva, V. V., & Vasilchenko, Y. M., (2014). The Calculating Experiment for Metal/Carbon Nanocomposites Synthesis with the Application Avrami Equation.*Nanostructure, Nanomaterials and Nanotechnologies to Nanoindustry*(pp. 105–118, 436).Toronto–New Jersey: Apple Academic Press.
6. Wang, J. Q., Wu, W. H., & Feng, D. M., (1992). *The Introduction to Electron Spectroscopy* (XPS/XAES/UPS) (p. 640). Beijing: National Defense Industry Press.

CHAPTER 4

SIMULATION OF THERMAL FIELDS AND FORMATION OF DROPS AT WELDING OF MICROSYSTEMS

S. V. SUVOROV[1] and ALEXANDER V. VAKHRUSHEV[1,2]

[1]*Department of Mechanics of Nanostructures, Institute of Mechanics, Udmurt Federal Research Center, Ural Division, Russian Academy of Sciences, Izhevsk, Russia*

[2]*Department of Nanotechnology and Microsystems, Technic Kalashnikov Izhevsk State Technical University, Izhevsk, Russia, E-mail: vakhrushev-a@yandex.ru*

ABSTRACT

The study is devoted to numerical modeling of the electrode melting process under the influence of an electric arc on it in the process of micro-welding. The mathematical model of thermal conductivity in the electrode, taking into account the assumptions made, is applicable to all parts of the electrode. The values of the voltage and the strength of the welding current, which determine the value of the heat flux from the electric arc to the electrode, varied over a wide range. The researches allowed establishing the basic laws of heating and melting of the electrode.

4.1 INTRODUCTION

When creating micromechanical products, various types of welding are often used, including electric arc welding, called micro-welding. Micro welding is characterized by small diameters of the electrodes used and low welding current forces. It should also be noted that micro-welding has a high degree of automation.

Micro-welding is a complex phenomenon and involves various physical and chemical processes. Understanding these processes allows you to create reliable microsystems with the required parameters. In this regard, mathematical modeling of the entire set of processes accompanying micro-welding is highly relevant and is used to study various welding technologies and solve current technological issues.

There are several main problems:calculation of thermal fields, the study of the heating of the electrode and the welded elements of the micro-system, the melting of the electrode, the drop formation processes, the formation of a micro weld, the cooling of the system, and the formation of residual stresses.

The change in temperature fields during welding was studied in Refs. [1, 2] by the finite element method.The joint evolution of temperature fields and the stress-strain state of the welding area were modeled in the three-dimensional setting in Ref. [3].Numerical simulation of the drop formation processes during electric welding in paper [4] was performed. The transient thermo-mechanical (coupled) analysis of temperature and residual stress distributions on welded plates was performed in Ref. [5]. The aim of this work is the complex modeling of micro-welding processes within the framework of the concept of multi-level mathematical modeling of nanosystems [6].

This chapter presents the results of the first part of this task-simulation of thermal fields and drop formation processes at micro-welding on macro-level modeling.

4.2 FORMULATION OF THE PROBLEM

The design of most welds involves a gap between the parts to be welded and, as a consequence, the volume of the gap must be filled with additional metal. The source of metal can be either a filler material or an electrode, such electrodes are classified as melting, in both cases, it is usually a wire.

The task will be considered on the example of a melting electrode. In the process of welding, an electric arc is established between the parts to be welded and the electrode, which heats both the part and the electrode.

In general, the process of melting the electrode during welding can be divided into two stages:

1. When the electrode is heated, the moment comes when the temperature in the electrode reaches the melting temperature (T_m), and the internal energy (H) stored in it becomes equal to the heat of fusion (ΔH_m). When two of these conditions are fulfilled, the process of melting (drop formation) begins in the contact zone of the electrode with the electric arc, while the temperature in the electrode continues to increase until the temperature in it reaches the boiling point (T_b); and
2. When the boiling point (T_b) in the electrode is reached and the internal energy accumulates in it equal to the heat of boiling (ΔH_b), the material from which the electrode is made boils, which causes it to evaporate. At the moment of the beginning of the boiling of the electrode material, the dimensions of the melted portion of the electrode, drops, will no longer change. Denote the increment of the internal energy – ΔH.

When modeling the electrode melting process, we introduce the following assumptions:

- There is no heat exchange between the protective gas and the electrode;
- Thermal expansion of the electrode material is absent;
- The heat flux generated by the electric arc is evenly distributed over the end of the electrode;
- In the molten part of the electrode, the Reynolds number tends to zero (Re0);
- The calculation is carried out until the boiling of the electrode material.

The assumption of a uniform distribution of the heat flux over the end of the electrode follows from the fact that the electrodes used in welding micromechanical products have a small diameter.

Taking into account the assumptions about the absence of heat exchange between the electrode and the protective gas and the uniform distribution of the heat flux over the end of the electrode, the process of thermal conductivity in the electrode itself can be described in a one-dimensional formulation.

The process of heating the electrode is described by the following mathematical model of thermal conductivity:

$$\rho\frac{\partial T}{\partial \tau}=\frac{\partial}{\partial x}\left(k\frac{\partial T}{\partial x}\right)+\gamma\cdot j^2-\dot{m}_m\Delta H_m-\dot{m}_b\Delta H_b,\ 0\le x\le l_\infty \quad (1)$$

where c, ρ, k, γ are the specific heat capacity, density, thermal conductivity, electrical resistivity of the electrode material, respectively; T is the temperature of the electrode material; τ is time; x is the spatial coordinate; j is the density of the electric current flowing through the electrode; $\dot{m}_m$, $\dot{m}_b$, are the specific mass rate of melting and boiling, respectively; l_∞ is the distance from the end of the electrode to the coordinate where there is no influence from the heating of the electrode by the electric arc.

The assumption that the Reynolds number in the molten portion of the electrode tends to zero leads to the fact that the process of convective heat transfer in the molten portion as a whole does not affect the process of heat conduction in the electrode.

The boundary and initial conditions, taking into account the assumptions introduced, will be determined by the following relations:

$$\left.\begin{aligned}&-k\frac{\partial T(0,\tau)}{\partial x}=q,\ 0\le\tau\le\tau_1,\\&-k\frac{\partial T(l_\infty,\tau)}{\partial x}=0,\ 0\le\tau\le\tau_1,\\&T(x,0)=293\ K,\end{aligned}\right\} \quad (2)$$

where q is the heat flux generated by the welding arc; τ_1 – the moment of the beginning of the boiling of the electrode material. The processes of melting and boiling the material of the electrode were simulated in accordance with the following dependencies [7]:

$$\left.\begin{aligned}
&\rho\left(\partial Y_m/\partial \tau\right)=\dot{m}_m;\ \dot{m}_m=\left(c\rho\Delta T_m\right)/\Delta H_m;\\
&\Delta T_m=\begin{cases}0,\ if\ \ T(\tau+\Delta\tau)<T_m;\\ \left[T(\tau+\Delta\tau)-T_{mn}(\tau)\right]/\Delta\tau\ if\ \ T(\tau+\Delta\tau)>T_m;\end{cases}\\
&T_{mn}(\tau)=\max\left(T(\tau),T_m\right);\\
&\rho\left(\partial Y_b/\partial \tau\right)=\dot{m}_b;\ \dot{m}_b=\left(c\rho\Delta T_b\right)/\Delta H_b;\\
&\Delta T_b=\begin{cases}0,\ if\ \ T(\tau+\Delta\tau)<T_b;\\ \left[T(\tau+\Delta\tau)-T_{bn}(\tau)\right]/\Delta\tau\ if\ \ T(\tau+\Delta\tau)>T_b;\end{cases}\\
&T_{bn}(\tau)=\max\left(T(\tau),T_b\right);
\end{aligned}\right\}\qquad(3)$$

where $\Delta\tau$ is the time step; Y_m, Y_b, are the mass fractions of the electrode material, melted, and boiled, respectively.

Low-carbon steel is considered as an electrode material. In this case, in accordance with [8, 9] and considering the fact that the parameters of low-carbon steels are close to iron [10], the thermophysical properties of the electrode material are taken equal to:

- = 7770 kg/m^3;k = 44,4 J/(m·sec·K);
- c = 557 J/(kg·K); T_{m} = 1800 K;
- ΔH_{m} = 247100 J/kg; T_{b} = 3145 K;
- ΔH_{b} = 6267123 J/kg; γ = 1.5·10^{-7}Ohm·m.

The current density is determined by the following relationship:

$$\left.\begin{aligned}
&j=\frac{I}{S},\\
&S=0,25\,\pi\,d^2,
\end{aligned}\right\},\qquad(4)$$

where I is the current strength of the welding arc; S is the cross-sectional area of the electrode; d is the diameter of the electrode.

The heat flux uniformly distributed over the end of the electrode is defined as [8]:

$$\left.\begin{aligned} q &= \frac{\lambda}{\pi}\eta UI, \\ \lambda &= 4\,d^{-2}, \end{aligned}\right\}, \qquad (5)$$

where η – the efficiency of conversion of electric power of the welding arc into heat; λ – concentration ratio of the welding arc; U is the arc voltage.

The type of electrode used is metal, in which case [11] the efficiency is assumed to be $\eta = 0.7$. The diameter of the electrode is assumed to be $d = 2.5 \cdot 10^{-4}$ m, in this case, $\lambda = 6.4 \cdot 10^{7}$ m^{-2}. We will accept the following ranges of welding arc parameters:

- *Umin* = 30 V – the minimum voltage of the welding arc;
- *Umax*= 60 V – maximum voltage of the welding arc;
- *Imin* = 0.225 A – minimum current of the welding arc;
- *Imax* = 0.625 A – the maximum current of the welding arc.

Table 4.1 shows the values of the heat flux of the welding arc, calculated from the relation (4.5) for the four combinations of voltage and the current strength of the welding arc.

TABLE 4.1 Heat Flow Welding Arc

Heat Flow of Welding Arc	$\frac{\lambda}{\pi}\eta U_{\min} I_{\min}$	$\frac{\lambda}{\pi}\eta U_{\max} I_{\min}$	$\frac{\lambda}{\pi}\eta U_{\min} I_{\max}$	$\frac{\lambda}{\pi}\eta U_{\max} I_{\max}$
q, J/(m^2·sec)	$9{,}626 \cdot 10^{7}$	$1{,}925 \cdot 10^{8}$	$2{,}674 \cdot 10^{8}$	$5{,}348 \cdot 10^{8}$

The problem of heat conduction (4.1) with initial and boundary conditions (4.2) is solved using the method of control volumes [12, 13]. The discrete analog of the mathematical model (4.1) is implemented in an explicit scheme; as a result, the time step is limited from above:

$$\Delta\tau < \frac{\tilde{n}\,\rho\,\Delta x_{\min}^{2}}{2\,k}, \qquad (6)$$

where $\Delta x_{\min}$ is the smallest size of the control volume in the entire computational domain. The calculation is carried out with the following parameters of the calculated area: – the size of the control volume is: $\Delta x = 10^{-5}$ m.

The control volume with the smallest size is "half," that is:

$$\Delta x_{\min} = 0.5\Delta x \tag{7}$$

From Eq. (7) it follows that $\Delta x_{\min} = 5 \cdot 10^{-6}$ m.

Substituting the value of $\Delta x_{\min}$ in (4.6), we determine that the time step should not exceed: $\Delta\tau < 1.218 \cdot 10^{-6}$ sec. The time step is taken equal to $\Delta\tau = 10^{-6}$ sec. The size of the computational domain determined by the value l_∞ is assumed to be equal, $l_\infty = 15 \cdot 10^{3}$ m.

4.3 RESULTS OF MODELING

A graphical representation of the simulation results is shown in Figures 4.1–4.4.

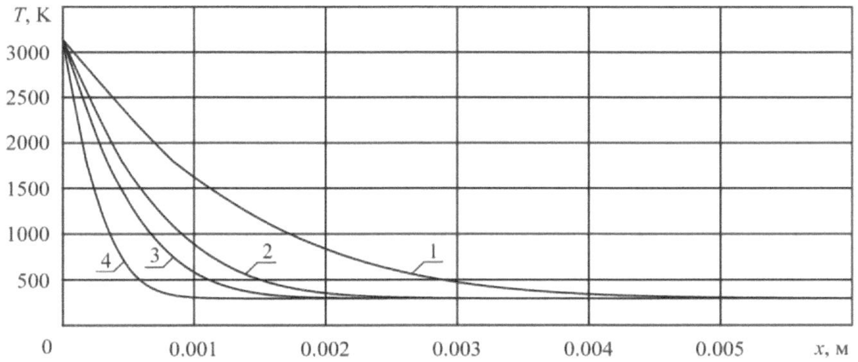

FIGURE 4.1 Temperature profiles along the length of the electrode at different values of the heat flux welding arc: 1 – $q = 9{,}626 \cdot 10^7$, 2 – $q = 1{,}925 \cdot 10^8$, 3 – $q = 2{,}674 \cdot 10^8$, 4 – $q = 5{,}348 \cdot 10^8$, J/(m^2·sec)

Graphs in Figure 4.1 show that the temperature gradient in the electrode is proportional to the heat flux entering the end of the electrode, therefore, as the heat flux decreases, the depth by which the electrode warms up increases, and at $q = 9{,}626 \cdot 10^7$J/(m^2·sec)it reaches the order value 7 mm.

The size of the electrode melt zone varies inversely with the heat flux from 0.84 mm at q = $9.626 \cdot 10^7$ J/(m^2·sec)to 0.18 mm at q = $5.348 \cdot 10^8$ J/(m^2·sec).

This is shown in Figure 4.2. In addition, in the place of the electrode where the temperature reaches the value of 1800 K, the temperature gradient changes, which is associated with the expenditure of energy on the accumulation of heat of fusion in the material.

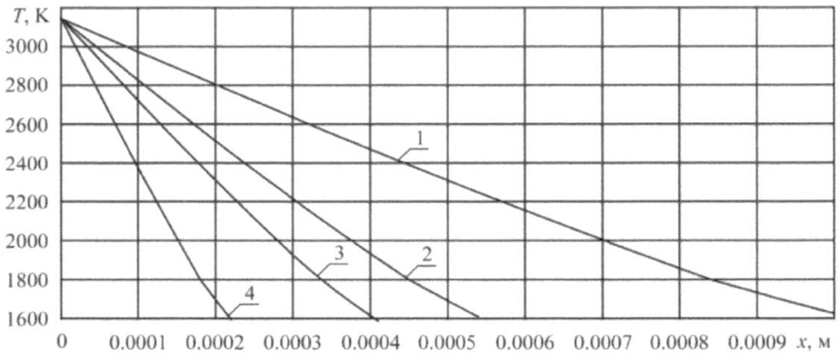

FIGURE 4.2 Temperature profiles in the zone of molten electrode material at different values of the heat flux welding arc:1 $-q = 9{,}626 \cdot 10^7$, 2 $-q = 1{,}925 \cdot 10^8$, 3 $-q = 2{,}674 \cdot 10^8$, 4 $-q = 5{,}348 \cdot 10^8$, J/(m^2·sec).

As can be seen from the graphs of Figure 4.3, when the temperature reaches 1800 K and 3145 K on the surface of the end of the electrode, the temperature stops rising. Graphically, this is displayed as horizontal sections, until the heat of melting and boiling is accumulated, respectively. It also follows from the graphs that, with a constant heat flux from the moment the melting point reaches the electrode material, the rate of temperature rises decreases.

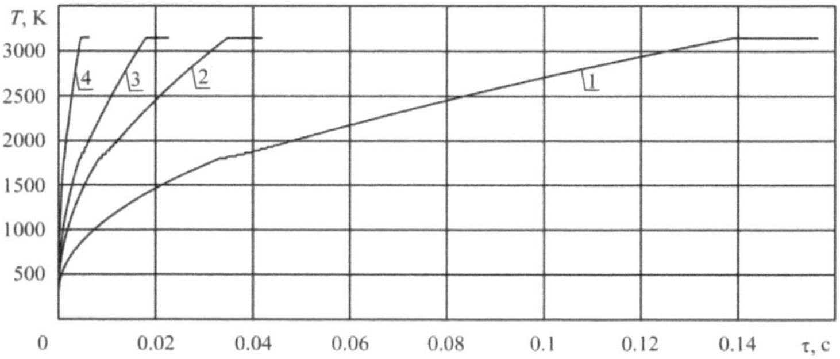

FIGURE 4.3 The temperature change on the surface of the end of the electrode at different values of the heat flux welding arc:1 – $q = 9{,}626 \cdot 10^7$, 2 – $q = 1{,}925 \cdot 10^8$, 3 – $q = 2{,}674 \cdot 10^8$, 4 – $q = 5{,}348 \cdot 10^8$, J/(m^2·sec).

The graphs shown in Figure 4.4, characterize the change in the heating time of the electrode from the magnitude of the heat flux.

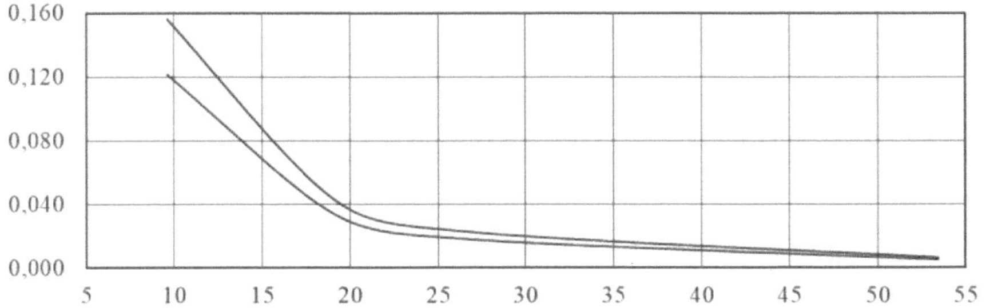

FIGURE 4.4 Effects of heat flux on the heating time of the electrode: 1 – total heating time before the boiling of the electrode material; 2 – time of formation of the electrode material.

If the curve 1 in Figure 4.4 shows the total heating time on the surface of the electrode end from 293 K (initial temperature) until the boiling point of the electrode material begins, then curve 2 represents the time of drop formation. This is the time from the moment of melting to the beginning of the boiling of the electrode material. From the graphs in Figure 4.4, it also follows that the dependence of the boiling time of the electrode material and the time of its drop formation on the heat flux generated by the welding arc is hyperbolic in nature.

The numerical values of time spent on boiling and dropping of the electrode material are given in Table 4.2

TABLE 4.2 The Start Time of Melting and Boiling of the Electrode Material

	$q = 9{,}626 \cdot 10^7$, J/(m^2·sec)	$q = 1{,}925 \cdot 10^8$, J/(m^2·sec)	$q = 2{,}674 \cdot 10^8$, J/(m^2·sec)	$q = 5{,}348 \cdot 10^8$, J/(m^2·sec)
Warming up time before the boiling of the electrode material, sec	0.15640	0.0418	0.0226	0.0062
Droplet time of the electrode material, sec	0.122	0.033	0.018	0.005
The proportion of the time of electrode drop in the total warm-up time,%	78.00	78.95	79.65	80.64

The values given in Table 4.2 show that when the heat flux entering the electrode from the welding arc changes in the studied range, the proportion of time spent on the electrode drop in the total heating time remains very constant and is about 79%.

4.4 CONCLUSIONS

The conducted studies, considering the assumptions made, made it possible to establish the main parameters of the electrode melting process during electric arc welding of micromechanical products.So, the influence of the heat flux on the temperature gradient in the electrode and the depth of the molten section are multidirectional.

A link was also established between the change in the heat flux and the time parameters of the drop formation process.In general, it can be noted that the accounting of energy costs in the process of electric arc welding on the accumulation of heat of melting and boiling makes a significant contribution to the overall melting pattern of the electrode. The studies performed will allow investigating in more detail the various processes during micro-welding at different structural levels.

ACKNOWLEDGMENTS

The works was carried out with financial support from the Research Program of the Ural Branch of the Russian Academy of Sciences (project 18-10-1-29) and budget financing on the topic "Experimental studies and multi-level mathematical modeling using the methods of quantum chemistry, molecular dynamics, mesodynamics, and continuum mechanics of the processes of formation of surface nanostructured elements and metamaterials based on them" (project 0427-2019-0029).

KEYWORDS

- **boiling**
- **drop formation**
- **electrode**
- **micro-welding**
- **numerical simulation**
- **thermal conductivity**

REFERENCES

1. Ali, M., (2012). Finite-element simulation for thermal profile in shielded metal arc welding (SMAW) process. *International Journalof Emerging Trends in Engineering and Development*, *1*(2), 9–18.
2. Janakiram, G., Vijay, S., &Venkateswara, R. M., (2012). Analysis of temperature distribution of different welded joints in ship building. *International Journal of Engineering Research and Technology*, *1*(6), 1–4.
3. Li, C., & Wang, Y., (2013). Three-dimensional finite element analysis of temperature and stress distributions for in-service welding process.*Materials and Design*, *52*, 1052–1057.
4. Suvorov, S. V., &Vakhrushev, A. V., (2018). Numerical modeling of the process of dropship of electrode at welding.*Chemical Physics and Mezoscopy*, *20*(3), 335–341.
5. Swapnil, R. D., Sachin, P. A., &Awanikumar, (2015). Finite element analysis of residual stresses on ferritic stainless-steel using shield metal arc welding.*International Journal of Engineering Research and General Science*, *3*(2), 1131–1137.
6. Vakhrushev, A. V., (2017). *Computational Multiscale Modeling of Multiphase Nanosystems: Theory and Applications*. Apple Academic Press: Waretown, New Jersey, USA.
7. Lipanov, A. M., Makarov, S. S., Karpov, A. I., &Makarova, E. V., (2017). Simulation study of a hot metal cylinder cooling by gas-liquid flow.*Thermo Physics and Aeromechanics*, *24*(1), 53–60.
8. Krektuleva, R. A., &Batranin, A. V., (2012). The joint solution of the inverse heat conduction problem and optimal design problems in the technology of non-consumable electrode welding.*Proceedings of Tomsk Polytechnic University*, *2*, 104–109.
9. Kazantsev, E. I., (1975). *Industrial Furnaces*. Moscow: Metallurgiya Publisher.
10. Babichev, A. P., Babushkina, N. A., &Bratkovskiy, A. M., (1991). *Physical Quantities*. Moscow: EnergoatomizdatPublisher.
11. Vasilev, K. V., &Vill,' V. I., (1979). *Welding in Mechanical Engineering*. Moscow: MashinostroeniePublisher.
12. Patankar, S. V., (1980). *Numerical Heat Transfer and Fluid Flow*.New York, Hemisphere Publisher.
13. Samarskiy, A. A., (1971). *Introduction to the Theory of Difference Schemes*. Moscow: NaukaPublisher.

CHAPTER 5

EFFECT OF IMPOSING LINEAR PARTICLE GRADIENT ON CREEP BEHAVIOR IN COMPOSITE DISC HAVING HYPERBOLIC THICKNESS

VANDANA GUPTA[1] and SATYA BIR SINGH[2]

[1]*Department of Mathematics, Dashmesh Khalsa College, Zirakpur (Mohali), Punjab – 160059, India*

[2]*Department of Mathematics, Punjabi University, Patiala – 147004, India*

ABSTRACT

In the present work, an effort has been made to study the effect of variation in linear particle gradient (PG) on the steady-state creep behavior in anisotropic disc made of functionally graded material disc (FGM) with hyperbolic thickness. In the anisotropic FGM disc, the content of silicon carbide particles decreases linearly from the inner radius to the outer radius of the disc. The creep response of the anisotropic disc under stresses developing due to rotation at 15,000 rpm has been determined by Sherby's law. The creep parameters of the anisotropic FGM disc vary along the radial distance due to varying composition. The creep behavior of anisotropic disc is expressed by a threshold stress with the value of stress exponent as 8. The study reveals that in the FGM disc, the radial stress increases throughout the disc with an increase in PG, whereas the stresses (tangential and effective) increase near the inner radius but decrease near the outer radius. By employing higher PG in anisotropic FGM disc with hyperbolic thickness, the distribution of steady-state strain rates becomes more uniform compared to an anisotropic disc having uniform reinforcement distribution. Thus, the care to introduce PG in anisotropic FGM disc should be taken for an optimal design of composite disc.

5.1 INTRODUCTION

Nowadays, advancement in technology has made it possible to synthesize materials for components that exhibit graded-variation in their properties. Under severe thermomechanical loadings, the conventional materials (metals or ceramics) may not survive alone. Thus, a new material concept of functionally graded material was introduced and led to the development of superior heat resistant materials. The idea of FGM was conceived to provide a material that may withstand severe thermomechanical loadings requiring heat-resistant ceramics on the high-temperature side and tough metals with high thermal conductivity on the lower temperature side. This variation in composition from ceramic to metal is made gradually in between the two sides are characterized in a way that the composition of each one and the volume fraction of materials are changed gradually. By gradually varying the volume fraction of reinforcement, their material properties exhibit a smooth and continuous change from one surface to another. The smooth variation of properties always offers many advantages such as the reduction of thermal stress, increased strength, etc. A major advantage of FGMs is the possibility of designing its gradation to optimize its performance. While designing the material properties, we need to determine the material phase volume fraction of each point of the structure. However, several authors have studied the stress and strain in the non-FGM/FGM discs made of isotropic/anisotropic material at elevated temperature using different yield criteria and creep law. Wahl et al. [29] were the first to investigate theoretically steady-state creep behavior by a power function in a rotating turbine disc made of 12% chromium steel using von Mises and Tresca yield criteria theoretically describing creep behavior and compared the results with experimental values. Ma [15] derived the formulas based on the maximum shear theory associated with the von Mises flow rule for calculating the results for the distributions of stresses and strain rates creep in variable thickness disc rotating at uniform temperature by using the power function for creep behavior. The results obtained by using the von Mises criterion are found to be in excellent agreement with the available experimental creep data

[29]. Ma [15] extended his work for rotating disc having a variable thickness, used in gas turbine and jet engine which are used in aircraft power plants and the nuclear space auxiliary power system. The study was based on the theory of the Tresca criterion. It is concluded that the distributions of stresses over the central portion of the variable thickness disc are quite different from the constant thickness disc. Bhatnagar et al. [2] investigated the creep response of orthotropic variable thickness discs by describing creep behavior with Norton's power law. Thickness of rotating disc can be constant, linear, and hyperbolic. It is concluded that stresses and strain rates in composite disc can be reduced by selecting an optimum profile of thickness and a certain type of anisotropy. Mishra and Panday [16] proposed that creep response in rotating composite disc made by aluminum alloy can be described in a better way by Sherby's constitutive creep model, as compared to Norton's creep model. Pandey et al. [18] studied the steady-state creep response in aluminum-based composites consisting of silicon carbide particles under uniaxial loading conditions with variation in the temperature range between 623K and 723K for a different combination of particle sizes (1.7 *μm*, 14.5 *μm* and 45.9 *μm*) with varying particle content (10 *vol*%, 20 *vol*% and 30 *vol*%). It is concluded that the composite with finer particle size has better creep resistance than that containing coarser ones. Durodola and Attia [7] investigated the benefits of using different forms of fiber gradation in rotating hollow and solid FGM discs with constant thickness. It is noticed that the stress and deformation distribution can be modified by the different forms of property gradation with the same nominal volume fraction of reinforcement modify in the FGM discs compared with uniformly reinforced discs. Singh and Ray [26] studied the creep analysis in an isotropic FGM rotating disc at uniform elevated temperature by using Norton's power law and concluded that the steady-state creep response in FGM disc is significantly superior compared to a non-FGM disc. Orcan and Eraslan [17] investigated that the stresses in composite discs with variable thickness are lower as compared to the results obtained for the disc having constant thickness than. Gupta et al. [12] have analyzed creep behavior of a rotating isotropic constant thickness disc made of FGM containing varying amounts of silicon carbide in the radial direction and operating in presence of radial thermal gradient and concluded that the steady-state strain rates in the rotating disc with the presence of thermal gradient and a linear particle gradient (PG) are significantly lower than that observed in an isotropic disc having uniform

distribution of particle content and operating under isothermal condition. Jahed et al. [13] observed that the use of variable thickness disc helps in minimizing the weight of disc which helps to reduce the overall payload in the aerospace industry. This implies that a disc with variable thickness has no restrictions on the limiting value of maximum disc speed compared to a disc with constant thickness. Rattan et al. [22] have analyzed steady-state creep response of an isotropic FGM disc with constant thickness by using Sherby's constitutive model. The results of isotropic disc having non-linear variation of particle distribution along the radial distance are compared with the discs having a uniform and linear distribution of particles along the radial distance. Hasan Callioglu et al. [4] studied that stress analysis on functionally graded rotating annular discs subjected to temperature distributions parabolically decreasing with radius. He concluded that the tangential stress by the increase in temperature decreases at the inner part of disc, but increases at the outer radius in disc whereas the radial stress reduces gradually for all distributions in the temperature. Garg et al. [9] studied the creep behavior analysis in variable thickness disc made of functionally graded material and noticed that the magnitudes of strain rates in functionally graded rotating discs are significantly lower than in a uniform composite disc. An FGM disc having reinforcement distributed in a nonlinear way possesses the lowest and relatively uniform distribution of strain rates. Deepak et al. [5] investigated the effect of thickness gradient on creep response in a linearly varying thickness disc made of functionally graded material containing silicon carbide particles in a matrix of pure aluminum. It can be concluded that the stresses and strain rates in both the radial as well as tangential directions reduce significantly with the increase in thickness gradient of the composite disc. Khanna et al. [14] analyzed creep behavior in a rotating isotropic disc with constant thickness by using the Tresca criterion. It is concluded that the gradient variation significantly effects the distribution of stresses and strain rates of the composite disc for purpose of designing a disc. Gupta and Singh [10, 27] studied an analytic framework for analyzing stresses and strain rates in isotropic rotating non-FGM/FGM disc with the thickness (constant, linear, and hyperbolic). The study revealed that the stresses and strain rates in the FGM hyperbolic thickness disc, are the lowest and more uniform compared to non-FGM/FGM discs with the thickness (constant and linear). Garg [8] studied the effect of varying thickness profile in the rotating FGM disc having linearly varying thickness. It is concluded that the strain rates are lower in the FGM disc having a higher thickness gradient along the radial direction. Bose and Rattan [3] have investigated the effect of the temperature gradient on anisotropic disc having a constant thickness. The results for the stress and strain rate distributions under a graded temperature field are expressed graphically. It is concluded that the effect of the parabolic temperature gradient should be considered while designing the anisotropic rotating disc due to its effect on the Disc's creep behavior.

S. B. Singh and coworkers have also analyzed Elastic-Plastic and creep transition in the disc using transition theory where the assumptions: (i) incompressibility condition, (ii) creep-strain laws like Norton, (iii) yield condition like that of Tresca, (iv) associated flow rule were not considered. The necessity of the use of ad-hoc semi-empirical laws in the classical theory of elastic-plastic transition is based on the approach that the transition is a linear phenomenon which is not possible. Under the elastic-plastic and creep transition, the fundamental structure of the object undergoes a change and rearrange themselves to cause non-linear effects. Therefore, it suggests that at transition behavior, non-linear terms are significant and cannot be ignored. The generalized strain measures are useful to solve the various problems of elastic-plastic by solving the non-linear differential equations at the transition points. This concept of generalized strain measures and transition theory has been applied to find transitional stresses in various problems. Thakur [19] discusses the problems in creep transition stresses of a thick isotropic spherical shell by finitesimal deformation under a steady-state of temperature by using Seth transition theory. All these problems based on the recognition of the transition state as a separate state necessitates showing the existence of the constitutive equation for that state. Thakur et al. [20, 21] further studied elastic-plastic and creep deformation in a rotating disc subjected to parameters such as variable thickness, variable density, etc.

The whole above literature study is based on pertaining to creep in rotating composite disc. With these forethoughts, it is decided to investigate the effect of imposing various kinds of linear PG on steady-state creep behavior for non-FGM/FGM disc. The material of the anisotropic disc is made by aluminum alloy based metal matrix composites containing silicon carbide particulates due to the excellent mechanical properties like high specific strength/stiffness and high-temperature stability. The thickness of the composite is assumed to be hyperbolic, because the stresses produced in the disc are due to rotary motion, can be minimized by varying the thickness of the disc. The analysis has been done by using Hill's criterion for yielding. The creep behavior of the composite disc with stress exponent 8 has been described by Sherby's constitutive law. The creep parameters in the law have been determined using the regression equations developed on the basis of available experimental results in the literature.

5.2 DISTRIBUTION OF REINFORCEMENT AND DISC PROFILE

In the present study, an anisotropic non-FGM/FGM disc of density ρ is considered with rotating with constant angular speed ω radian/sec. The inner (a) and outer (b) radii of all the discs are considered, respectively. The FGM disc has silicon carbide particles varying linearly from inner to outer radius. Therefore, the creep parameters and density of the composite disc will vary with radial distance. The material properties of the annular disc are assumed to be functions of the volume fraction of the constituent materials. The composition variation in terms of volume percent of silicon carbide, along the radial distance, $V(r)$, is given as

$$V(r)=V_{\max}-\frac{(r-a)}{(b-a)}\left(V_{\max}-V_{\min}\right) \quad (1)$$

$$V(r)=A-Br, \quad a\leq r\leq b \quad (2)$$

where,

$$A=V_{\max}+\mathrm{a}\,B \quad (3)$$

and

$$B=\frac{V_{\max}-V_{\min}}{b^2-a^2} \quad (4)$$

where, $V_{\max}$ and $V_{\min}$, are the maximum and minimum particle contents, respectively, at the inner and outer radii of the disc.

According to the law of mixture, the density variation $\rho(r)$ in the composite disc may be expressed as:

$$\rho(r) = \rho_m + (\rho_d - \rho_m)\frac{V(r)}{100} \tag{5}$$

where, ρ_m (= 2713 kg/m^3) and ρ_d (= 3210 kg/m^3) are the densities of the pure aluminum (Al) matrix and of the dispersed silicon carbide particles, respectively.

Substituting the value of $V(r)$ from Eq. (2) into Eq. (5):

$$\rho(r) = \rho_m + (\rho_d - \rho_m)\frac{A - Br}{100} = A_\rho - B_\rho .r \tag{6}$$

where,

$$A_\rho = \rho_m + (\rho_d - \rho_m)\frac{A}{100} \tag{7}$$

and

$$B_\rho = (\rho_d - \rho_m)\frac{B}{100} \tag{8}$$

The average particle content in the FGM disc (V_{avg}) may be expressed as:

$$V_{avg} = \frac{\int_a^b 2.\pi.r.h(r)\,V(r)d}{t(b^2 - a^2)\pi} \tag{9}$$

Since, the thickness $h(r)$ of the composite disc having hyperbolically varying thickness as shown in Figure 5.1, is assumed to be of the form,

$$h(r) = c\,r^m \tag{10}$$

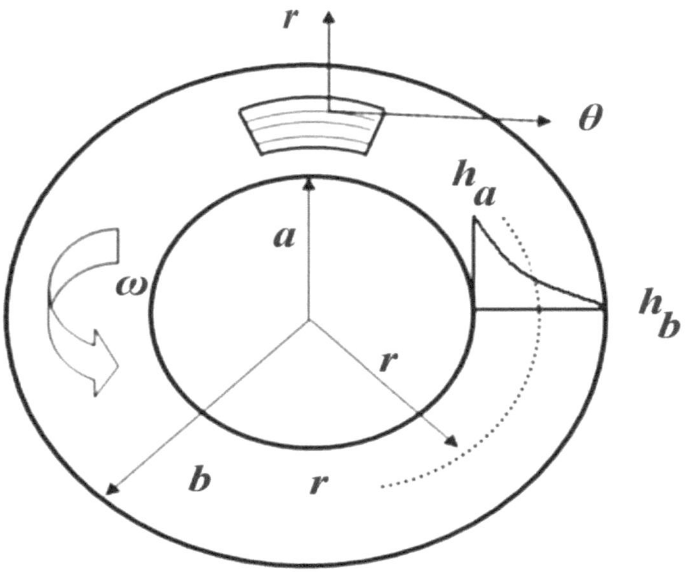

FIGURE 5.1 A rotating disc of hyperbolic thickness.

where, $c = 31.26$ and $m = -0.74$ are constants [2].

If the FGM disc is having hyperbolically varying thickness as given in Eq. (10), then by substituting the expressions of $h(r)$ and $V(r)$ from Eq. (10) and Eq. (1) into Eq. (9), V_{min} may be obtained as,

$$V_{min} = \frac{V_{max}\left[a^{m+2}\left(b\,(m+3) - a\,(m+2)\right) - b^{m+3}\right] + \frac{t}{2c}V_{avg}\left(b^2 - a^2\right)(b-a)(m+2)(m+3)}{b^{m+2}\left(b\,(m+2) - a\,(m+3)\right) + a^{m+3}} \quad (11)$$

where,

$$t = \frac{2\left(b^{m+2} - a^{m+2}\right)}{(m+2)(b^2 - a^2)}c$$

Let I and I_0 denote the moment of inertia of the disc at inner radius a and outer radius r and b, respectively. A and A_0 denote the area of cross section of disc at inner radius a and outer radius r and b, respectively. Then,

$$I = \int_a^r h\, r^2\, dr, \qquad I_0 = \int_a^b h\, r^2\, dr$$

$$A = \int_a^r h\, dr, \qquad A_0 = \int_a^b h\, dr \tag{12}$$

where, the average tangential stress ($\sigma_{\theta avg}$) may be defined as

$$\sigma_{\theta avg} = \frac{1}{A_0} \int_a^b h\, \sigma_\theta\, dr. \tag{13}$$

5.3 CREEP LAW AND ESTIMATION OF CREEP PARAMETERS

The composite rotating FGM disc made of 6061Al matrix composite containing silicon carbide particle is assumed to undergo steady-state creep following a well-documented threshold stress-based creep law, may be expressed as [23]:

$$\dot{\bar{\varepsilon}} = \left[M(r) \left(\bar{\sigma} - \sigma_0(r) \right) \right]^n \tag{14}$$

where,

$$M(r) = \frac{1}{E} \left(\frac{A D_\lambda \lambda^3}{|b_r|^5} \right)^{1/n} \quad \text{is the creep parameter.}$$

where, $\dot{\bar{\varepsilon}}$, $\bar{\sigma}, n$, σ_0, A, D_λ, λ, b_r, E are the effective strain rate, effective stress, the stress exponent, threshold stress, a constant, lattice diffusivity, the subgrain size, the magnitude of burgers vector, Young's modulus.

In particle reinforced composite, the creep parameters $M(r)$ and $\sigma_0(r)$ depend on the particle size and the percentage of dispersed particles apart from the temperature, have been obtained from the creep results by using the experimental results reported by Pandey et al. [18] for Al-SiCp composite under uniaxial loading are given in Table 5.1.

For the values of creep parameters, the regression analysis has been performed in DATAFIT software to estimate the values of $M(r)$ and $\sigma_0(r)$

in terms of P,T and $V(r)$. The developed regression equations are given below,

$$M(r) = e^{-35.38} P^{0.2077} T^{4.98} V(r)^{-0.622} \quad (15)$$

$$\sigma_0(r) = -0.03507P + 0.01057T + 1.00536V(r) - 2.11916 \quad (5.16)$$

In a FGM disc with silicon carbide particle content varying radially as $V(r)$, both the creep parameters $M(r)$ and $\sigma_0(r)$ will vary due to variation with particle content. In the present study, the particle size (P) and the operating temperature (T) are taken as 1.7 μm and 623 K over the entire disc. Thus, for a given FGM disc under a known PG both the creep parameters are functions of radial distance and their values $M(r)$ and $\sigma_0(r)$ at any radius (r), could be determined by substituting the values of particle and temperature into Eqs. (15) and (16), respectively.

TABLE 5.1 Creep Parameters Based on the Experimental Results

Particle Size P (μm)	Temperature T (K)	Particle Content V ($Vol\%$)	M ($s^{-1/8}$/MPa)	σ_0 (MPa)
1.7	623	10	0.00963	15.24
14.5			0.01444	11.46
45.9			0.01897	13.65
1.7	623	10	0.00963	15.24
		20	0.00594	24.83
		30	0.00518	34.32
1.7	623	20	0.00594	24.83
	673		0.00897	24.74
	723		0.01295	25.72

Source: Pandey et al. [18].

5.4 MATHEMATICAL FORMULATION OF CREEP IN FGM DISC

Consider an aluminum silicon-carbide particulate composite disc of varying thickness having an inner radius, a = 31.75 mm and outer radius, b = 152.4 mm rotating with angular velocity, ω = 15,000 rpm. For the purpose of analysis of the disc, the following assumptions are made:

1. Material of disc is incompressible and locally isotropic i.e., the properties of the disc remain constant at a given radius in all the directions but can change with the change in radius.
2. Elastic deformations are small for the disc and therefore they can be neglected as compared to creep deformation.
3. Stresses at any point of the disc remain constant with time i.e., steady-state condition of stress is assumed.
4. Axial stress in the disc may be assumed to be zero as the thickness of disc is assumed to be very small compared to its diameter.
5. The composite shows a steady-state creep behavior, which may be described by Sherby's law as given by Eq. (15).

Taking reference frame along the principal directions of r, θ and z, the generalized constitutive equations for creep in an anisotropic composite disc under multiaxial stress takes the following form:

$$\dot{\varepsilon}_r = \frac{\dot{\bar{\varepsilon}}}{2\bar{\sigma}}\{(G+H)\sigma_r - H\sigma_\theta - G\sigma_z\} \tag{17}$$

$$\dot{\varepsilon}_\theta = \frac{\dot{\bar{\varepsilon}}}{2\bar{\sigma}}\{(H+F)\sigma_\theta - F\sigma_z - H\sigma_r\} \tag{18}$$

$$\dot{\varepsilon}_z = \frac{\dot{\bar{\varepsilon}}}{2\bar{\sigma}}\{(F+G)\sigma_z - G\sigma_r - F\sigma_\theta\} \tag{19}$$

where, F, G, and H are anisotropic constants of the material. $\dot{\varepsilon}_r$, $\dot{\varepsilon}_\theta$, $\dot{\varepsilon}_z$ and σ_r, σ_θ, σ_z are the strain rates and the stresses, respectively, in the direction r, θ and z. $\dot{\bar{\varepsilon}}$ be the effective strain rate and $\bar{\sigma}$ be the effective stress. For biaxial state of stress σ_r, σ_θ, the effective stress [11] is:

$$\bar{\sigma} = \left\{ \frac{1}{\left(\frac{G}{F}+\frac{H}{F}\right)} \left\{ \sigma_\theta^2 + \frac{G}{F}\sigma_r^2 + \frac{H}{F}(\sigma_r - \sigma_\theta)^2 \right\} \right\}^{1/2} \tag{20}$$

Using Eqs. (7) and (20), Eq. (17) can be rewritten as,

$$\dot{\varepsilon}_r = \frac{d\dot{u}_r}{dr} = \frac{\left[\left(\frac{G}{F}+\frac{H}{F}\right)x(r) - \frac{H}{F}\right]\left[M(r)\ (\bar{\sigma}-\sigma_0(r))\right]^8}{\sqrt{\frac{G}{F}+\frac{H}{F}}\left[\left(\frac{G}{F}+\frac{H}{F}\right)x(r)^2 - 2\,\frac{H}{F}\,x(r) + \left(\frac{G}{F}+\frac{H}{F}\right)\right]^{1/2}} \tag{21}$$

Similarly from Eq. (18):

$$\dot{\varepsilon}_\theta = \frac{\dot{u}_r}{r} = \frac{\left[\left(1+\frac{H}{F}\right) - \frac{H}{F}x(r)\right]\left[M(r)\ (\bar{\sigma}-\sigma_0(r)\right]^8}{\sqrt{\frac{G}{F}+\frac{H}{F}}\left[\left(\frac{H}{F}+\frac{G}{F}\right)x(r)^2 - 2\,\frac{H}{F}\,x(r) + \left(1+\frac{H}{F}\right)\right]^{1/2}} \tag{22}$$

From the material's incompressibility assumption, it follows that:

$$\dot{\varepsilon}_z = -\left(\dot{\varepsilon}_r + \dot{\varepsilon}_\theta\right) \tag{23}$$

where, $x(r) = \frac{\sigma_r}{\sigma_\theta}$ is the ratio of radial and tangential stresses.

Dividing Eq. (21) by Eq. (22) and integrating the resulting equation by taking limit a to r on both sides:

$$\dot{u}_r = \dot{u}_{r_i}\ \exp\int_a^r \frac{\varphi(r)}{r}\,dr \tag{24}$$

where, $\dot{u}_{r_i}$, is the radial deformation rate at the inner radius, $\dot{u}_r = du/dt$ is the radial deformation rate and

$$\varphi(r) = \frac{\left(\frac{G}{F}+\frac{H}{F}\right)x(r) - \frac{H}{F}}{\left(1+\frac{H}{F}\right) - \frac{H}{F}x(r)}$$

Substituting $\dot{u}_r$ from Eq. (24) into Eq. (22), we get the tangential stress (σ_θ),

$$\sigma_\theta = \frac{(\dot{u}_{r_i})^{1/n}}{M(r)}\ \psi_1(r) + \psi_2(r) \tag{25}$$

where,

$$\psi_1(r) = \frac{\psi(r)}{\left\{\left(\frac{1}{\frac{G}{F}+\frac{H}{F}}\right)\left[\left(\frac{G}{F}+\frac{H}{F}\right)x(r)^2 - 2\frac{H}{F}x(r) + \left(1+\frac{H}{F}\right)\right]\right\}^{1/2}} \tag{26}$$

$$\psi_2(r) = \frac{\sigma_0(r)}{\left\{\left(\frac{1}{\frac{G}{F}+\frac{H}{F}}\right)\left[\left(\frac{G}{F}+\frac{H}{F}\right)x(r)^2 - 2\frac{H}{F}x(r) + \left(1+\frac{H}{F}\right)\right]\right\}^{1/2}} \tag{27}$$

and

$$\psi(\mathrm{r}) = \left\{\frac{\sqrt{\frac{G}{F}+\frac{H}{F}}}{r} \cdot \frac{\left[\left(\frac{\mathrm{H}}{\mathrm{F}}+\frac{\mathrm{G}}{\mathrm{F}}\right)x(r)^2 - \frac{2\mathrm{H}\,x(r)}{\mathrm{F}} + \left(1+\frac{\mathrm{H}}{\mathrm{F}}\right)\right]^{1/2}}{\left[\left(1+\frac{\mathrm{H}}{\mathrm{F}}\right) - \frac{\mathrm{H}}{\mathrm{F}}x(r)\right]} \exp.\int_{\mathrm{a}}^{\mathrm{r}} \frac{\varphi(\mathrm{r})\mathrm{dr}}{\mathrm{r}}\right\}^{1/8} \tag{28}$$

The equation of equilibrium for a rotating disc with varying thickness can be written as,

$$\frac{d}{dr}\left(r\; h(r)\; \sigma_r\right) - h(r).\sigma_\theta + \rho(r)\omega^2 r^2 h(r) = 0 \tag{29}$$

where, $\rho(r)$ is the density of FGM disc.

Integrating Eq. (29) within limits a to b and substituting the values of $\rho(r)$ and $h(r)$

$$\sigma_{\theta_{avg}} = \frac{\omega^2}{A_0}\left[A_\rho I_0 - B_\rho.c.\left(\frac{b^{m+4} - a^{m+4}}{m+4}\right)\right] \tag{30}$$

where $\sigma_{\theta_{avg}}$ denote the average tangential stress of FGM disc with hyperbolic thickness.

Substituting σ_θ from Eq. (25)

$$\left(\dot{u}_{r_i}\right)^{1/n} = \frac{A_0\ \sigma_{\theta_{avg}} - \int_a^b \psi_2(r).\, h(r)\, dr}{\int_a^b \frac{\psi_1(r).\, h(r)}{M(r)} dr} \tag{31}$$

Using Eq. (31), Eq. (25) becomes,

$$\sigma_\theta = \frac{\psi_1(r)\left[A_0\ \sigma_{\theta_{avg}} - \int_a^b \psi_2(r).\, h(r)\, dr\right]}{M(r)\int_a^b \frac{\psi_1(r).\, h(r)}{M(r)} dr} + \psi_2(r) \tag{32}$$

Integrating Eq. (29) within limits *a* to *b* and substituting the values of $\rho(r)$ and $h(r)$

$$\sigma_r = \frac{1}{r.h(r)}\left[\int_a^r \sigma_\theta .h(r)\, dr - \omega^2 A_\rho I + \omega^2 B_\rho\, c.\left(\frac{r^{m+4} - a^{m+4}}{m+4}\right)\right] \tag{33}$$

Thus, for FGM disc with hyperbolic thickness, the tangential stress σ_θ and radial stress σ_r are determined by Eqs. (32) and (33), respectively. Then strain rates $\dot{\varepsilon}_r$, $\dot{\varepsilon}_\theta$ and $\dot{\varepsilon}_z$ calculated from Eqs. (21), (22), and (23).

5.5 NUMERICAL COMPUTATIONS

The distributions of stress and strain rate are evaluated for the non-FGM/FGM disc with hyperbolic thickness from the analysis procedure described in section 4, through an iterative numerical scheme of computation shown in Figure 5.2. The iteration is continued till the convergence of the process and gives the results of stresses at different points of the radius grid. For rapid convergence 75% of the value of σ_θ obtained in the current iteration has been mixed with 25% of the value of σ_θ obtained in the last iteration for use in the next iteration, i.e., $\sigma_{\theta next} = .25\sigma_{\theta previous} + .75\sigma_{\theta current}$.

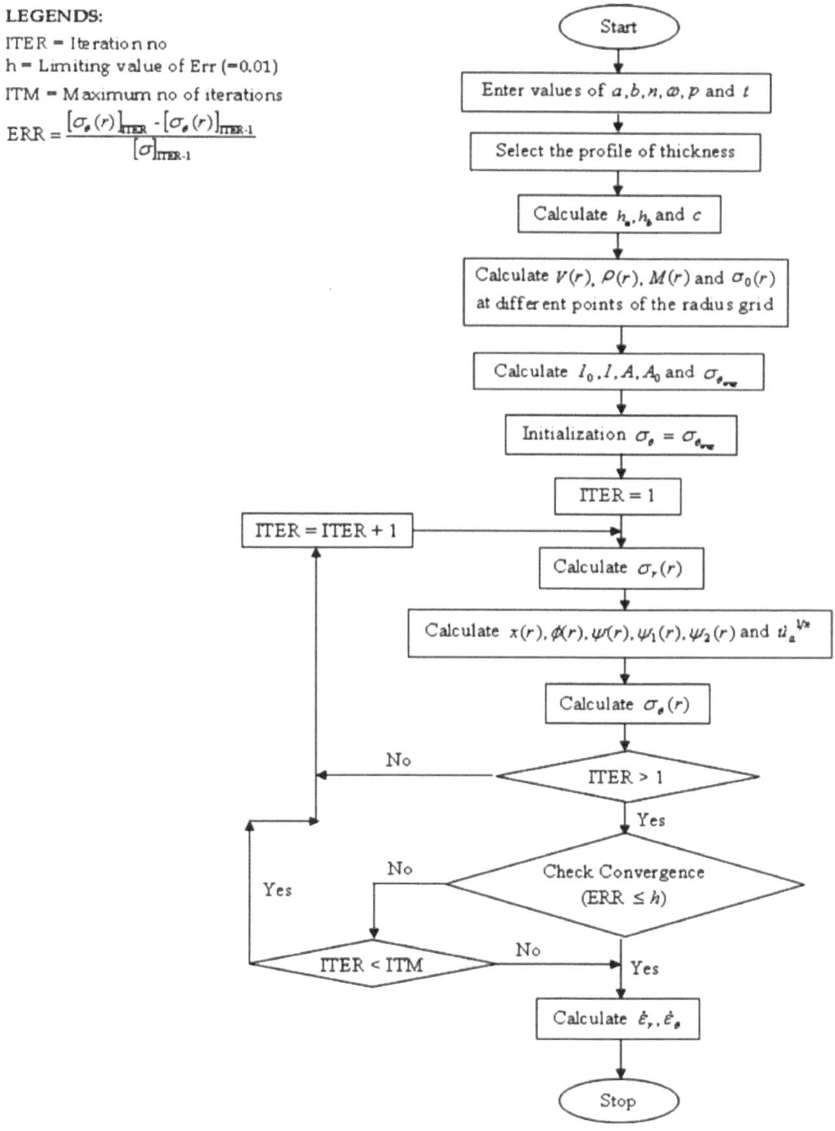

FIGURE 5.2 Numerical scheme of computation.

5.6 RESULTS AND DISCUSSION

A computer code based on the mathematical formulation presented in this paper has been developed to obtain stresses and creep rates in the rotating non-FGM/FGM disc with same average content of silicon carbide particles. For all the FGM discs, maximum particle content V_{max} and minimum particle content V_{min} is taken from Table 5.3. After this, the content of silicon carbide particle, $V(r)$ in the different FGM discs is calculated from Eq. (1). Thus, obtained values of $V(r)$, for all FGM discs with hyperbolic thickness are substituted in Eqs. (15) and (16) to get the variation of the creep parameters, $M(r)$ and $\sigma_0(r)$.

5.6.1 VALIDATION

Before discussing the results of the steady-state creep response obtained in this study, it is considered necessary to validate the analysis carried out and the software developed. To achieve this goal, the results for a rotating composite disc by following the current analysis scheme were obtained for the disc with operating conditions for which are mentioned in Table 5.2.

TABLE 5.2 Parameters and Operating Conditions for Disc

Parameters for steel disc:
For non-FGM disc, density of disc material $\rho = 2862.1 kg/m^3$
Density, Al = 2713 kg/m^3
Density, SiC = 3210 kg/m^3
Inner radius of disc, a = 31.75 mm
Outer radius of disc, b = 152.4 mm
Particle size, P = 1.7 μm
Uniformly distributed particle content, V_{avg} = 20%
For non-FGM disc creep parameters: $M(r)$ = 5.94×10^{-3} $s^{-1/8}/MPa$ and $\sigma_0(r)$ = 24.83 MPa
Young's modulus, Al = 70 GPa
Young's modulus, SiC = 47 GPa
Anisotropic constants, $\frac{G}{F}=1.34$, $\frac{H}{F}=1.64$
The disc thickness for hyperbolically varying thickness at the inner radii, h_a = 2.42 mm

TABLE 5.2 *(Continued)*

The disc thickness for hyperbolically varying thickness at the outer radii, $h_b = 0.76\ mm$
Operating conditions:
Angular velocity of Disc, $\omega = 15{,}000\ rpm$
Operating temperature, $T = 623\ K$
Creep duration, $t = 180\ hrs$

Figure 5.3 shows the distribution of reinforcement in various composite discs. The particle content $V(r)$ in FGM disc with hyperbolic thickness decreases linearly from the inner to the outer radius and in non-FGM disc, the particle content is uniform (20 vol%) over the entire radius. Although, the variation of $V(r)$ becomes steeper in the anisotropic FGM disc (D2).

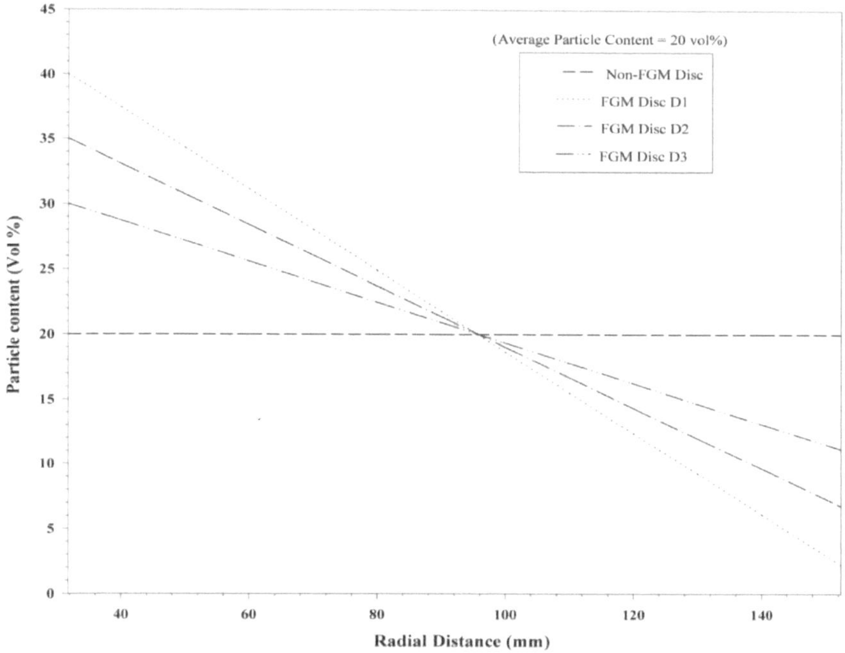

FIGURE 5.3 Variation of particle content in anisotropic non-FGM/FGM discs having hyperbolic thickness.

Figures 5.4 and 5.5 shows the variation of material parameters *M*(*r*) and $\sigma_0(r)$, respectively, along radial distance for the different (FGM and non-FGM) composite discs. Both the material parameters, *M*(*r*) and $\sigma_0(r)$ observed to be constant over the entire non-FGM disc due to uniform distribution of particle content of silicon carbide (20 vol%) in aluminum matrix. But in FGM disc, the creep parameter *M*(*r*) increases

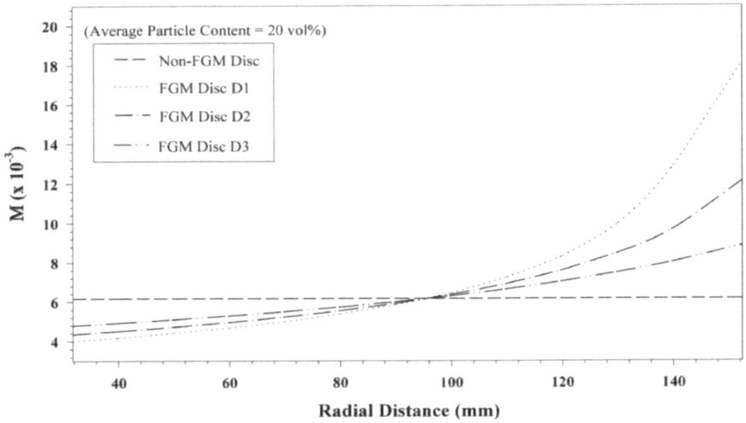

FIGURE 5.4 Variation of creep parameter *M* in anisotropic non-FGM/FGM discs having hyperbolic thickness.

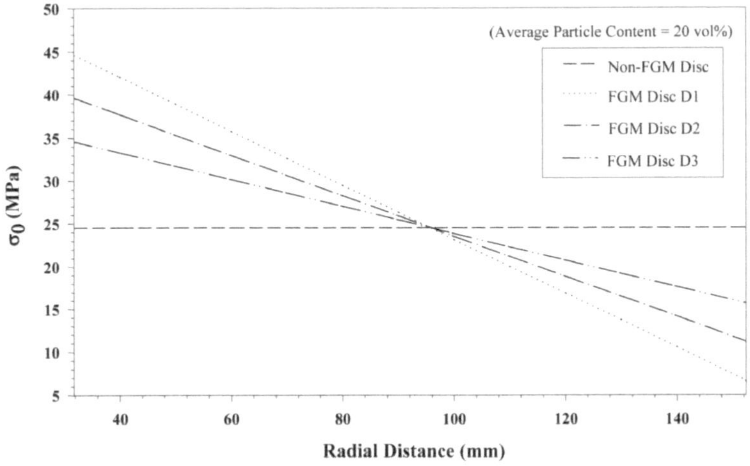

FIGURE 5.5 Variation of creep parameter S_0 in anisotropic non-FGM/FGM discs having hyperbolic thickness.

with increasing the radial distance due to decrease in particle content $V(r)$ on moving from the inner to the outer radius, as evident from Eq. (14). It is also observed that the creep parameter $M(r)$ becomes much steeper with increase in PG beyond 19%. On the other side, the threshold stress $\sigma_0(r)$ in FGM disc decreases linearly on moving from the inner to the outer radius. The threshold stress is maximum at inner radius in composite (FGM disc D2) because of more amount of particle content of silicon carbide, but near outer radius, the threshold stress is minimum shown in Figure 5.5.

TABLE 5.3 Description of Rotating Anisotropic Non-FGM/FGM Discs of Variable Thickness

Disc (Notation)	Particle (SiCp) Content (vol%)			Particle Gradient (PG) = $V_{max}-V_{min}$ (vol%)
	V_{max}	V_{min}	V_{avg}	
Uniform/Non-FGM (D1)	20	20	0	20
FGM (D2)	40	2.23	20	37.77
FGM (D3)	35	6.64	20	28.36
FGM (D4)	30	11.16	20	18.84

5.6.2 EFFECT OF LINEAR DISTRIBUTION OF PARTICLE CONTENT ON CREEP BEHAVIOR OF ROTATING HYPERBOLICALLY VARYING THICKNESS DISC

Figures 5.9 to 5.12 show the effect of PG on the steady-state creep behavior of the non-FGM and FGM discs with hyperbolically varying thickness. The material of all the composite discs (non-FGM disc and FGM) is anisotropic. It is observed that the tangential stress in all the composite discs increases with increasing radial distance, reaches maximum and then decreases on moving toward the outer radius of the discs as shown in Figure 5.8. The FGM discs with hyperbolic thickness which having linearly decreasing particle content from the inner to outer radius as shown in Figure 5.3, has a relatively higher tangential stress near the inner radius but lower values of tangential stress near the outer radius than that observed in non-FGM discs with hyperbolic thickness. The primary reason for these changes in the tangential stress distribution depends on the ratio of the value of $\psi(\mathrm{r})$ at a given radial distance to its average value and multiplied by the average value of $\sigma_{\theta_{avg}}$ which is dependent on density, as shown in Eq. (32). The average value of $\sigma_{\theta_{avg}}$, does not depend on radial distance and so the variation of tangential stress will reflect only variation of σ_{θ}. It is observed that distribution of $\psi(r)$ in FGM disc D2 is higher near inner radii and lower in outer radii compared to other FGM/non-FGM discs (D1, D3, D4) as shown in Figure 5.7. On the other hand, in Figure 5.6, the higher density near the inner radius due to higher amount of particle content and relatively lower density near the outer radius due to small amount of particle content have resulted in a higher and lower tangential stress, respectively, compared to tangential stress in the non-FGM disc having the average density due to uniform distribution of particles. The trend of variation of tangential stress is similar in all non-FGM/FGM discs. The tangential stress in anisotropic rotating FGM disc D2 with PG (PG=37.77%) having hyperbolic thickness is reduced near outer radii compared to other FGM/non-FGM disc (D1, D3, D4). It is also observed that the variation of effective stress with radial distance for non-FGM disc/FGM disc as shown in Figure 5.8, is similar as variation of tangential stress.

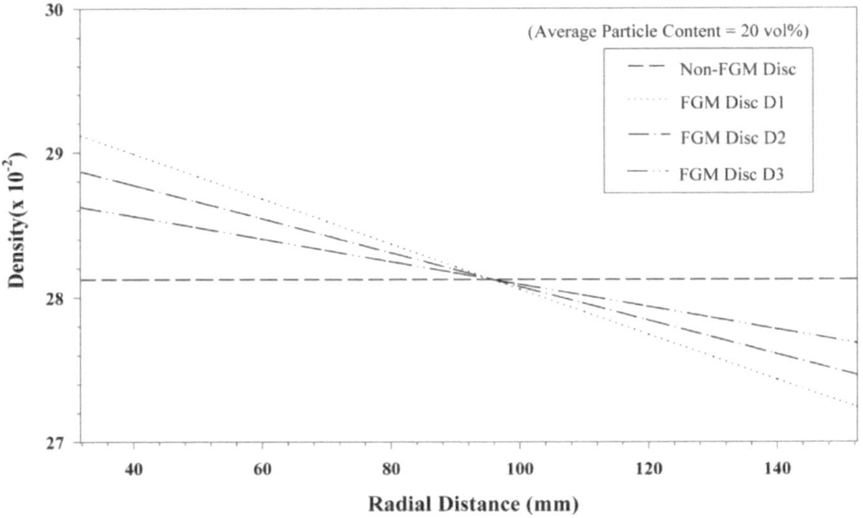

FIGURE 5.6 Variation of density in anisotropic non-FGM/FGM discs having hyperbolic thickness.

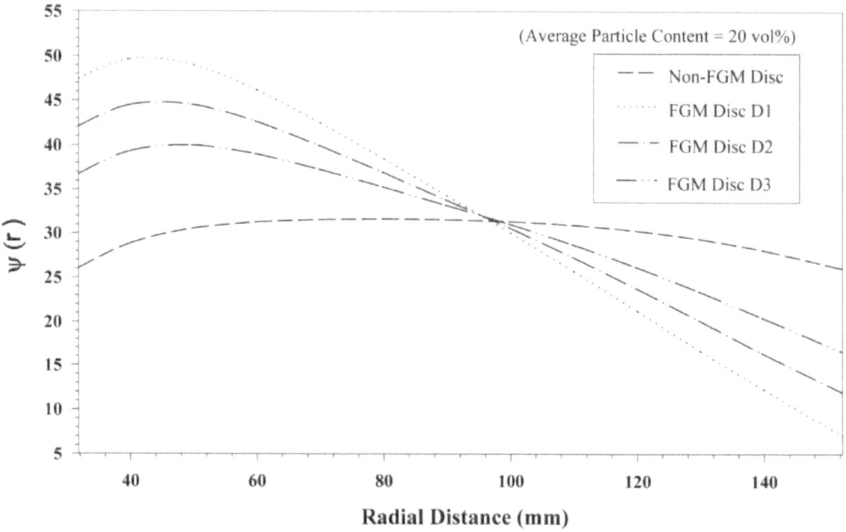

FIGURE 5.7 Variation of function $\psi(\mathrm{r})$ in anisotropic non-FGM/FGM discs having hyperbolic thickness.

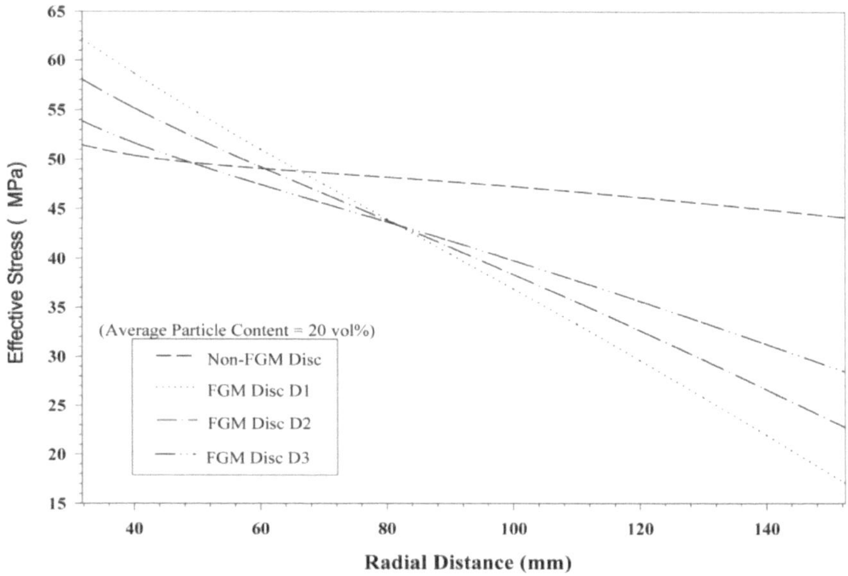

FIGURE 5.8 Variation of effective stress in anisotropic non-FGM/FGM discs having hyperbolic thickness.

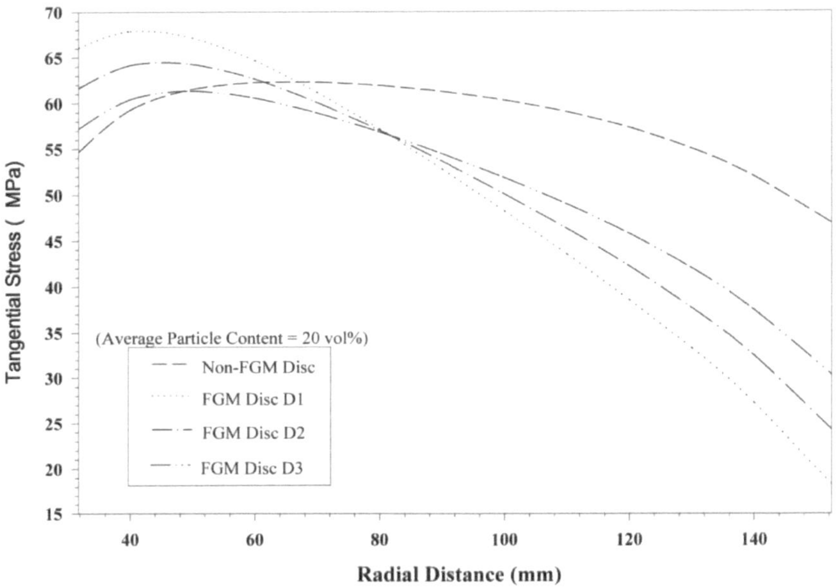

FIGURE 5.9 Variation of tangential stress along the radial distance in anisotropic non-FGM/FGM discs having hyperbolic thickness rotating with an angular velocity 15,000 rpm at 623K.

The radial stress increases from zero to reach a maximum value in the middle of the non-FGM and FGM disc followed by a decrease to reach at zero level again at the outer radius, under the imposed boundary conditions of vanishing radial stress at the inner and the outer radii of the disc, as shown in Figure 5.10. The radial stress is highest over the entire disc by increasing PG in FGM disc and lowest in the non-FGM disc which having uniform distribution of particle content.

In all the FGM/non-FGM discs with hyperbolic thickness, the tangential strain rate is highest at the inner disc and then decreases continuously when one moves towards the outer disc as shown in Figure 5.11. Secondly, the tangential strain rate in FGM discs is lower about two to three order of magnitude than the non-FGM disc (D1). It is also evident that the steady-state tangential strain rate developing in a rotating FGM hyperbolic thickness disc (D2) having higher PG is relatively more uniform. This uniformity in distribution, in the region near the inner radius, is due to varying particle content in FGM disc, whereas near the outer radius, it is due to lower tangential stress in spite of lower particle content, resulting due to lower density of the FGM disc. The effect of PG on radial strain rate is similar to tangential strain rate as shown in Figure 5.12. By increasing PG beyond 37% in FGM disc, the distribution of radial strain rate becomes more uniform compared to other non-FGM/FGM discs and the nature of radial strain rate, which is generally compressive in FGM disc (D1 and D4), becomes tensile in middle of the FGM disc (D2 and D3). The FGM disc D2 possesses higher PG, have maximum tensile radial strain rate in middle part of Disc. Thus, the radial strain rate in FGM disc (D2) is reduced significantly by about three to four order of magnitude compared to non-FGM disc (D1). Therefore, a disc having higher PG (D2) will have less chance of distortion.

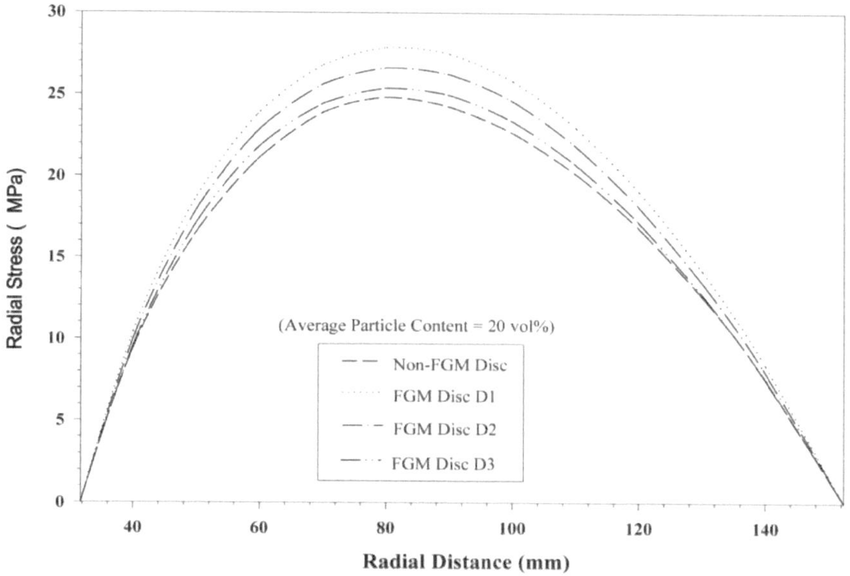

FIGURE 5.10 Variation of radial stress along the radial distance in anisotropic non-FGM/FGM discs having hyperbolic thickness rotating with an angular velocity 15,000 rpm at 623K.

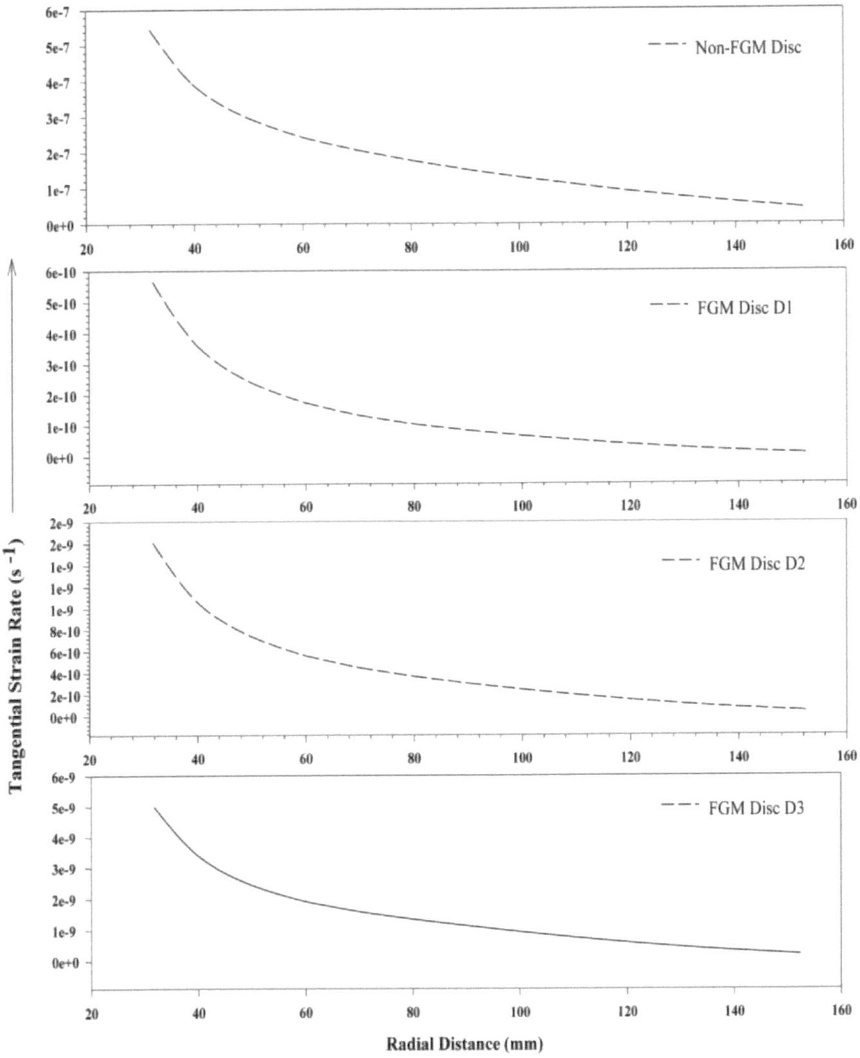

FIGURE 5.11 Variation of tangential strain rate along the radial distance in anisotropic non-FGM/FGM discs having hyperbolic thickness rotating with an angular velocity 15,000 rpm at 623K.

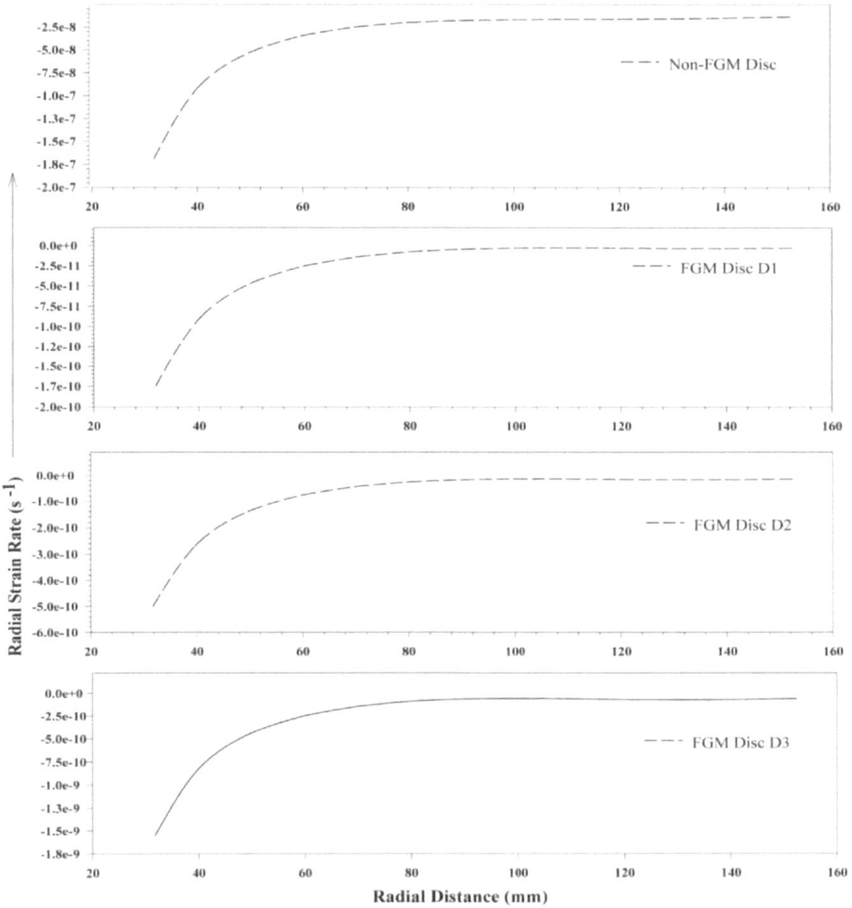

FIGURE 5.12 Variation of radial strain rate along the radial distance in anisotropic non-FGM/FGM discs having hyperbolic thickness rotating with an angular velocity 15,000 rpm at 623K.

5.7 CONCLUSION

The above results and discussion concludes that:

1. The distribution of tangential and radial stress in the anisotropic disc with hyperbolic thickness is affected by varying the variation of PG. The magnitude of tangential stress near the outer radius

of the FGM disc having higher PG is lower than the disc having uniform or lower PG. However, the effect of PG on the tangential stress towards the inner radius in composite disc is opposite to that observed near the outer radius.

2. By employing higher PG in the anisotropic rotating FGM disc with hyperbolic thickness, the magnitude of tangential creep rate can be reduced by about three orders of magnitude in compare to disc having uniform PG.
3. In FGM disc with linearly decreasing particle content from the inner to the outer radius, the distribution of radial strain rates becomes relatively uniform as compared to a disc containing uniform distribution of reinforcement, which may reduce the chances of distortion in the disc.
4. The distribution of reinforcement helps in restraining the creep in the rotating composite disc with variable thickness. Thus, for designing an anisotropic FGM disc with hyperbolic thickness, the variation of PG needs attention from the point of view of creep behavior of disc.

KEYWORDS

- **functionally graded material**
- **particle gradient**
- **radial strain rate**
- **steady-state creep**
- **tangential creep rate**
- **tangential stress**

REFERENCES

1. Arya, V. K., & Bhatnagar, N. S., (1979). "Creep analysis of rotating orthotropic disc." *Nuclear Engineering and Design, 55*, 323–330.
2. Bhatnagar, N. S., Kulkarni, P. S., & Arya, V. K., (1986). "Steady-state creeps of orthotropic rotating discs of variable thickness." *Nuclear Engineering and Design, 91*(2), 121–141.

3. Bose, T., & Rattan, M., (2018). "Modeling creep analysis of thermally graded anisotropic rotating composite disc." *International Journal of Applied Mechanics 10*(6), 1850063.
4. Callioglu, H., Demir, E., & Sayer, M., (2011). "Thermal stress analysis of functionally graded rotating discs." *Scientific Research and Essays, 6*(6), 3437–3446.
5. Deepak, D., Gupta, V. K., & Dham, A. K., (2013). "Investigating the effect of thickness profile of a rotating functionally graded disc on its creep behavior." *Journal of Thermoplastic Composite Materials, 26*(4), 461–475.
6. Deepak, D., Gupta, V. K., & Dham, A. K., (2010). "Mathematical modeling of steady-state creep in a rotating disc of variable thickness." *International Journal of Computational Material Science and Surface Engineering, 4*(2), 109–129.
7. Durodola, J. F., & Attia, O., (2000). "Deformation and stresses in functionally graded rotating discs." *Composites Science and Technology, 60*(7), 987–995.
8. Garg, M., (2017). "Stress analysis of variable thickness rotating FG disc." *International Journal of Pure and Applied Physics, 13*(1), 158–161.
9. Garg, M., Salaria, B. S., & Gupta, V. K., (2012). "Analysis of steady-state creeps in a functionally graded rotating disc of variable thickness." *Composites: Mechanics, Computations, Applications, an International Journal, 3*(2), 171–188.
10. Gupta, V., & Singh, S. B., (2016). "Mathematical modeling of creep in a functionally graded rotating disc with varying thickness." *Regenerative Engineering and Translational Medicine* (pp. 126–140) (Springer) 10.1007/s40883-016-0018-3.
11. Gupta, V. K., Singh, S. B., Chandrawat, H. N., & Ray, S., (2004). "Steady-state creep and material parameters in a rotating disc of Al-SiCp composite." *European Journal of Mechanics A/Solids, 23*(3), 335–344.
12. Gupta, V. K., Singh, S. B., Chandrawat, H. N., & Ray, S., (2005). "Modeling of creep behavior of a rotating disc in presence of both composition and thermal gradients." *Journal of Engineering Materials and Technology, 127*(1), 97–105.
13. Jahed, H., Farshi, B., & Bidabadi, J., (2005). "Minimum weight design of inhomogeneous rotating discs." *International Journal of Pressure Vessels and Piping, 82*(1), 35–41.
14. Khanna, K., Gupta, V. K., & Nigam, S. P., (2015). "Creep analysis of a variable thickness rotating FGM disc using Tresca criterion." *Defense Science Journal, 65*(2), 163–170.
15. Ma, B. M., (1960). "Creep analysis for rotating solid discs of variable thickness." *Journal Franklin Institute, 269*(5), 408–419.
16. Mishra, R. S., & Pandey, A. B., (1990). "Some observations on the high-temperature creep behavior of (6061). Al-SiC composites." *Metallurgical Transactions, 21A*(7), 2089–2091.
17. Orcan, Y., & Eraslan, A. N., (2002). "Elastic-plastic stresses in linearly hardening rotating solid discs of variable thickness." *Mechanics Research Communications, 29*(4), 269–281.
18. Pandey, A. B., Mishra, R. S., & Mahajan, Y. R., (1992). "Steady state creep behavior of silicon carbide particulate reinforced aluminum composites." *Acta Metallurgica Materialia, 40*(8), 2045–2052.

19. Pankaj, T., Sethi, M., Shivdev, S., Singh, S. B., & Emmanuel, F. S., (2018). "Exact solution of rotating disc with shaft problem in the elastoplastic state of stress having variable density and thickness." *Structural Integrity and Life, 18*(2), 126–132.
20. Pankaj, T., Sethi, M., Shivdev, S., Singh, S. B., Emmanuel, F. S., & Lozanović, Š. J., (2018). "Modeling of creep behavior of a rotating disc in the presence of load and variable thickness by using Seth transition theory." *Structural Integrity and Life, 18*(2), 133–140.
21. Pankaj, T., Shivdev, S., Nishi, G., & Singh, S. B., (2017). "Effect of mechanical load and thickness profile on creep in a rotating disc by using Seth's transition theory." *American Institute of Physics: Conference Proceedings (USA), 1859*, 20–24.
22. Rattan, M., Chamoli, N., & Singh, S. B., (2010). "Creep analysis of an isotropic functionally graded rotating disc." *International Journal of Contemporary Mathematical Sciences, 5*(9), 419–431.
23. Sherby, O. D., Klundt, R. H., & Miller, A. K., (1977). "Flow stress, subgrain size and subgrain stability at elevated temperature." *Metallurgical Transactions, 8A*, 843–850.
24. Shivdev, S., Singh, S. B., & Pankaj, T., (2019). "Modeling creep parameter in rotating discs with rigid shaft exhibiting transversely isotropic and isotropic material behavior." *The Journal of Emerging Technologies and Innovative Research, 6*(1), 387–395.
25. Singh, S. B., (2008). "One parameter model for creep in a whisker reinforced anisotropic rotating disc of Al-SiCw composite." *European Journal of Mechanics A/ Solids, 27*(4), 680–690.
26. Singh, S. B., & Ray, S., (2001). "Steady-state creep behavior in an isotropic functionally graded material rotating disc of Al-SiC composite." *Metallurgical Transactions, 32A*(7), 1679–1685.
27. Gupta, V., & Singh, S. B., (2017). "Impact of stress exponent on creep behavior in a rotating composite disc with hyperbolic thickness." *Journal for Technology of Plasticity, Belgrade, Serbia, 42*(1), 33–48.
28. Von Mises, R., (1913). "Mechanics of solids in the plastically deformable state." *NASA, Technical Memorandum, 88488*, 1986.
29. Wahl, A. M., Sankey, G. O., Manjoine, M. J., & Shoemaker, E., (1954). "Creep tests of rotating risks at elevated temperature and comparison with theory." *Journal of Applied Mechanics, 76*, 225–235.

CHAPTER 6

MICROSTRUCTURE AND PROPERTIES OF FIRE-RESISTANT POLYMERIC MATERIALS

S. G. SHUKLIN,[1,2] D. S. SHUKLIN,[1] and ALEXANDER V. VAKHRUSHEV[1,3]

[1]*Kalashnikov Izhevsk State Technical University, Izhevsk, Russia, E-mail: shuklin_sg@mail.ru*

[2]*Udmurt State University, Izhevsk, Russia*

[3]*Institute of Mechanics, Ural Division, Russian Academy of Sciences, Izhevsk, Russia*

ABSTRACT

From the analysis of the literature data, it follows that in the study of composite polymeric materials, special attention is not required. Changes in the chemical structure, the effect on the production process of components and components, as well as the heating conditions, have not yet been studied. The limited number of papers devoted to these problems is explained by the complexity of the processes under consideration.

6.1 POSSIBILITIES OF REDUCING THE FLAMMABILITY OF POLYMERS

The problem of reducing the flammability of polymeric materials is usually considered in the light of ideas about the multistage nature of the process of their diffusion burning. The slowdown and inhibition of the gross process can be achieved by actively affecting each stage by physical and chemical means [1], as well as by the method of mixed exposure.

Among the physical measures of influence on the combustion process are [2–4]:

1. Slowing down the supply of heat to the polymer material (for example, heat-insulating shielding of its surface);
2. Cooling of combustion zones as a result of an increase in physical heat sinks to the environment (for example, heat outflow from a polymer coating through a heat-conducting substrate, evaporation losses of components, heat loss by melted drops);
3. Deterioration of the conditions of transfer of reagents to the combustion front (creation of a physical barrier between the polymer and the oxidizing medium, slowing down the diffusion of combustible components in composites);
4. Disruption of the flame by the gas flow;
5. The impact of acoustic, gravitational, and other fields [1];
6. Processing of materials or finished products by thermal action, radiation with the creation of pressure and other conditions conducive to compaction and coking of the surface layers.

The process of chemical modification can include:

1. Modification of the polymer matrix by introducing fragments, oligomers, and/or monomers, leading to the formation of coke surface layers or chemical particles that inhibit the combustion processes in the pre-flame zone;
2. The introduction of active additives (reactive oligomeric or polymeric additives that are incompatible or partially compatible with the polymer matrix);
3. The introduction of active fillers that form hydrogen coordination bonds with the polymer matrix;
4. Surface modification (grafting of active particles onto the surface, forming blanks of the coke surface layer or inert gas layer in the pre-flame zone);
5. The introduction of inert additives:
    a. with the aim of subsequent separation of diluents of combustible gases;
    b. leading to the dispersion of filler particles or polymer degradation products from the surface;

c. for surface modification (surface treatment with oligomeric and polymeric substances that form interpenetrating or penetrating networks in the surface layer, increasing the likelihood of coking of the material) [5].

There are also mixed methods of action (physicochemical) – chemical modification of the polymer and the creation of a foam-coke layer using a heat pulse – which effectively reduces the flammability of polymeric materials [6, 7].

Physical and chemical measures of influence on the burning polymer system, carried out from the outside, are used in practice to suppress an already arisen and developing process, that is, to extinguish a fire. However, the actual problem of reducing the flammability of polymeric materials is associated with the use of such measures as prophylactic, preventing the possibility of occurrence or slowing down the development of burning materials. In other words, the influence of physical and chemical factors should manifest itself in the system itself, without external intervention.

In the response of the system to the effects of heat from an oxidizing environment or a fire, at some stage of the combustion process, "self-defense" physical or chemical factors may prevail. In this case, we can talk about the leading mechanism for reducing the flammability of a polymeric material.

6.2 EVALUATION OF THE EFFECTIVENESS OF FLAME RETARDANTS AND SYNERGISTIC FIRE RETARDANT SYSTEMS (FRS)

The efficiency of catalysts has been proposed [8] to be considered depending on the initial burning rate of known explosives: ammonium perchlorate, ammonium nitrate or nitroguanidine. Catalytic efficiency is represented as an equation:

$$K = AUon \quad (1)$$

The nature of the dependence is due to a sharp drop in catalytic activity with an increase in the burning rate according to the power-law above. In Ref. [9], in a certain approximation, the normalized burning rate is considered as an additive value of autocatalytic Ua and thermal Ur components, hence the factor of autocatalytic of the flame:

$$U_a = \frac{U_0}{U_1} \tag{2}$$

The same factor may be an indicator of inhibition of combustion. Such parameters are used under the action of catalysts or inhibitors of combustion in the pre-flame zone.

In the general case, the evaluation of the effectiveness of combustion regulators can be done in two ways: with respect to the concentration of combustion polymers chosen as the standard, and the combustion controller under study; by changing the combustion effect at the same concentrations of the standard and studied combustion regulator.

In most cases, especially when burning is inhibited, the concentration of the element or group that contributes to the combustion or cooking process is used to evaluate the effectiveness instead of the concentration of the combustion regulator. For example, to achieve the same damping effect of different polymers, the following concentrations of phosphorus, bromine, and chlorine are required [10] (Table 6.1).

TABLE 6.1 Evaluation of the Effectiveness of the Elements, %

Polymers	Phosphorus	Bromine	Chlorine
Polyolefins	5	–	40
Polyacrylates	5	16	20
Polyacrylonitriles	5	10–12	10–15
Polyurethanes	1,5	12–14	10–15
Polyethers	5	12–15	25
Epoxy resins	5–6	13–15	26–30

Thus, the most effective is phosphorus, then bromine, and then chlorine. If you choose for the standard phosphorus compounds in the material, then, by definition, the effectiveness evaluation can be represented as:

$$Э_{ph} = \left[C\right]st[C]mid \tag{3}$$

where [C]st is the concentration of the active element taken as the standard (in this case phosphorus); [C] mid – the concentration of the active element being compared, that is, the efficiency of phosphorus is taken to be unity, the efficiency of bromine varies from 0.5 to 0.31, and chlorine from 0.5 to 0.125.

In Ref. [55], the effectiveness of flame retardants was proposed to be assessed as a percentage:

$$EEf = \frac{\Delta Ef}{Ef_0}100 \quad (4)$$

where $\Delta Ef = Ef_0 - Ef_k$; here Ef_0 is the compound of the element necessary to achieve the extinction of materials based on polyesters and epoxy resins,% (Ef_0 on the basis of experimental data is 5.9%); Ef_k – connection of the element in a particular material, leading to the same effect,%.

The calculation can be made in fractions, since E takes negative values. As shown earlier, the combustibility of polymers depends on their chemical structure and composition, therefore, the assessment of the effectiveness of changes in flammability, and flammability is often associated with the chemical structure of combustion regulators. In particular, van Krevelen [11] proposed to use the ratio G = ΔOI/ (c + b), where G is the efficiency coefficient of combustion retardants, as an estimate of the effectiveness of flame retardants; ΔOI – change in oxygen index; c is the concentration of the element contributing to the reduction of flammability, %; b = 0.02 A (atomic mass of the element). The efficiency factors G of flame retardants are given for classes of flame retardants containing phosphorus, bromine, or chlorine (Table 6.2.).

TABLE 6.2 The Activity Coefficients of Some Classes of Flame Retardants for Various Polymers [11]

Class Retarders	Active Element	Values for Polymers*					
		PE	PS	PAN	PA	CL	PET
Phosphates	P	1. 3	–	1.5	1.1	1.5	1.5
Phosphates, phosphine oxides	P	–	1.5	–	–	–	–
Amidophosphates	P	–	–	–	–	2.4	2.6
Aromatic	CI	0.2	0.1	–	0.1	–	0.8
Without antimony oxide	Br	0.45	0.3	–	0.1	–	0.5
With antimony oxide	Br	1.0	–	–	–	–	–
Aliphatic	Br	0.6	1.1	0.35	–	–	0.7
Without antimony oxide	CI	0.2	0.5	0.1	0.1	–	0.8
With antimony oxide	CI	–	3.0	–	–	–	–

* PE – polyethylene, PS – polystyrene, PAN – polyacrylonitrile, PA – polyamide, CL – cellulose, PET – polyethylene terephthalate.

As can be seen from the table, the range of efficiency of flame retardants generally repeats the previously cited row, that is, phosphorus-containing substances are three to four times more efficient than bromine-containing ones, and they, in turn, are two to four times more effective than chlorine-containing flame retardants. Values for phosphorus, bromine, and chlorine are respectively 0.6; 1.6 and 0.7.

The van Crevelein equation [11] was derived for the limits of bromine and chlorine concentrations from 1 to 10% and phosphorus from 0.6 to 2.5%. Data analysis Table 6.2 indicates a higher activity of amid phosphates compared with phosphates when used in polymers that can form hydrogen bonds with additives, such as cellulose and polyethylene terephthalate (PET). These additives have less activity in polyamides, the macromolecules of which interact with each other. It is interesting to note that the dependence of KeCe + on the phosphorus compound in the range of 0.5–2.5% is linear, which coincides with the initial position of Van Crevelen in deriving the equation underlying the Table 6.2.

Thus, phosphorus compounds are characterized by interactions between their molecules and polymer macromolecules, which affect the effectiveness of these compounds as catalysts for carbonization processes.

At the same time, to evaluate the efficiency of combustion regulators, [9] use is made of the reaction of thermal oxidation of carbon-containing substances, such as diamond, graphite, and technical carbon. The authors suggest that substances interact on the surface with the corresponding tetrachloride elements with the formation of substances (monolayer on the surface), which can be represented as follows:

$$[C]_m (OH)_n \xrightarrow[-Hl]{+n \ni Cl_4} \cdot 4[C_m O_n \ni_n] Cl_{3-n}, m \square\ n$$

Catalytic effect is pronounced for compounds containing chromium and vanadium. The reaction rate of thermal oxidation for samples containing chromium increases by 203 times, vanadium – by 820 times. This reduces the activation energy of the process from 206 kJ/mol for an unmodified sample to 155 kJ/mol and 124 kJ/mol for chromium vanadium-containing samples of carbon black TG-10.

In the development of Eq. (1), we can present the formula of Wyle [10], which expresses the dependence of the oxygen index (CI) of polymers on the concentration of halogenated phosphates:

$$\Delta OI = R_p C_p + R_{b2} C_{b2} + R_{p1} B_2 C_p C_{b2}$$

and in general, based on statistical regression analysis:

$$\Delta OI = R_{b2} C_{b2}^{a} + R_p C_p + R_{p1} C_{p1} C_p C_{b2}$$

where Rp and Rp_1, Rb_2 are constants characterizing the efficiency of each active element entering the combustion controller and their joint action; Cp, Cb_2 is the concentration of active elements in the combustion controller.

According to the authors of work [12], the activation energy of the gaseous gross reaction can serve as a quantitative measure of the effectiveness of the flame retardant. The Eact value is calculated from the slope of the straight lines obtained in the coordinates: the relative value of the attenuation rate $\Delta \vartheta_e / \vartheta_e$ is the relative change in the concentration of oxygen in the current of the oxidant ($\Delta O_{ox}/O_{ox}$); or a relative change in its temperature ($\Delta T_0/T_0$) [12]. The authors distinguish a direct analogy of the considered method for evaluating the effectiveness of fire retardants, proposed earlier, with a measure of the effectiveness of inhibitors of the combustion of gas-phase systems [13]:

$$\Phi_v = \frac{[O_2]\Delta V}{[J]V}$$

where $[O_2]$ is the oxygen concentration; ΔV is the change in the rate of flame propagation with the introduction of a certain amount of inhibitor [J]; V is the flame propagation velocity in the absence of an inhibitor.

The interrelation of the chemical structure of polymers and thermal properties discovered by the authors is very interesting [14, 15]. The heat resistance of polymers prone to carbonization and the yield of carbonated residue during pyrolysis increase with an increase in the main chain of the macromolecule of aromatic groups. However, polycondensation heat-resistant polymers based on adducts of maleic anhydride and aromatic hydrocarbons are characterized by high heat resistance, but have low coke residue values and burn in the air due to the release of a large amount of combustible gases. Consequently, the linear relationship between heat resistance and coke residue yield is not always valid.

A promising method of increasing the efficiency of fire retardant systems (FRS) is the use of their mixtures. When the combined effect of two or more FRSs is greater than the sum of the effects of each of the individual components, a mixture of synergistic combustion retardation is said to occur, which may also arise as a result of the interaction of the additive with the hetero element already present in the polymer structure. Thus, an increase in the efficiency of phosphorus-containing FRS with their introduction into nitrogen-containing polymers was noted. It was shown that when phosphorus compounds are introduced into cellulose-containing nitrile groups, phosphorate synergy is manifested due to the formation of phosphoramide, which is an active dehydrating agent [15]. It is noted that nitrogen works as a flame retardant only in the presence of phosphorus in such compounds as methyloldimethylphosphonopropinamide and ammonium phosphorate [14].

To increase the versatility of FRS, it is promising to use variable-valence metal compounds as synergistic additives. The effectiveness of compounds of this type as FRS is shown in a number of works. For example, when adding 0.5% platinum to siloxane rubber, its acid index increases from 22 to 32, the output of coke increases by 50%, the vertically burning samples self-extinguish, while the initial samples burn completely [16]; similarly, palladium (Pd) and ruthenium (Re). Transition metal oxides significantly reduce the smoke emission during the combustion of polymers by increasing the oxidation rate of pyrolysis products, which leads to a decrease in the growth rate of soot particles [17]. Oxides of antimony, zinc, bismuth on organic halogen-containing carriers used as FRS for polyethylene (PE), polypropylene (PP), polyamide (PA), PET, polycarbonate (PC), polyurethane (PU), polystyrene (PS), epoxy resin (ES) [18]. Vanadium compounds can be very effective as additives to FRS. In organic chemistry, compounds of pentavalent vanadium are used as oxidation catalysts [19]. A mechanism has been proposed for the oxidation of polyatomic alcohols, including the formation of a complex of pentavalent vanadium with a substrate, followed by one-electron oxidation of alcohol in this complex by another molecule of vanadium [20].

Analysis of the literature shows that fire hazard is a total indicator that includes characteristics of flammability, loss of performance properties, characteristics of smoke generation, toxicity of gases and vapors. If any of the above is not taken into account when developing low-flammability composite materials, the proposed material will have limited use and will not be sufficiently effective. The use of fireproof materials increases the likelihood of resource conservation and improving the environmental safety of various industries in which these materials are used.

6.3 REGULATION OF THE PYROLYSIS PROCESS

A number of authors [1, 6, 21–23] one of the promising ways to reduce the flammability of polymeric materials is to regulate the process of pyrolysis, shifting its overall flow towards increasing the yield of carbonized residue. On the one hand, this reduces the yield of volatile combustible pyrolysis products, and on the other hand, the coke crust formed on the surface of the burning polymer shields the polymer from the heat flow of the flame, which reduces the surface temperature of the polymer material and makes it difficult for the flammable gases to escape. The appearance of a coke layer on the surface leads to a change in the heat and mass transfer processes on the surface of the burning polymer, and the heat flow to the non-destructive polymer decreases, which in turn leads to a change in the composition of the products in the gas phase – the effect of exfoliation during prolonged exposure.

Brauman [24] showed that, in the general case, the formation of coke leads to a decrease in flammability, burning rate, and flame spread rate, and the more coke-forming ability a polymer has, the more noticeable these effects are.

It was noted in Ref. [25] that the presence of a coke residue, which is formed during the destruction of a polymer under burning conditions, leads to a decrease in flammability and, accordingly, an increase in the oxygen index.

During the burning of polymers prone to carbonization, at the initial moment, a rapid decomposition of polymers is observed, and then as the carbonized residue accumulates on the surface, the rate of destruction decreases and a steady state is established [9, 26]. Such "stabilization," according to the authors of [26], suggests that along with the destruction, competing secondary processes occur at this stage associated with condensation, cross-linking, cyclization, aromatization, and the formation of thermally more stable polymer structures.

The surface of the burning polymer has a certain amount of oxygen, which can affect the processes in the surface layers of the polymer [9, 25], so the question of the participation of oxygen in the gasification process of the condensed phase is important for understanding the mechanism of combustion of polymers. In Ref. [9], the combustion and pyrolysis of high-impact polystyrene were investigated, where heat is supplied by radiation, and it was shown that the rate of linear pyrolysis does not depend on the presence of oxygen in the nitrogen-oxygen mixture. However, some authors believe that oxidative surface processes are the leading stage in the combustion of polymers. Thus, it has been shown [26] that oxygen unreacted in a flame diffuses from the gas zone to the surface of the condensed, being absorbed by the polymer and forming an oxygen-rich surface layer, in which the exothermic oxidative destruction of the polymer occurs.

In the case of polymers burning with the formation of coke, the burning rate of the carbonized layer obviously depends on how the oxidative reaction is performed: only on the surface of the carbonized layer or throughout its thickness in the presence of through pores.

To date, works devoted to the study of the influence of the morphological structure of coke on the combustion processes are clearly not enough.

The extinction of the polymer material is often due to the coking of the surface layer. In this case, coke can serve as a "screen" between the ignition source and unreacted material layers or a heat insulator (for example, foam cokes). Apparently, it is possible to present probable models of coke based on thermophysical parameters and density (Figure 6.1) [27].

The processes of coke formation and gas formation in the surface and subsurface layers of the material also determine the inhibition of combustion by dispersing with a drift the droplets of the formed particles of the liquid or solid phase. The ejection of these particles leads to a sharp decrease in temperature in local areas of the surface and a decrease in the overall energy of the surface layer of the material.

During combustion, thermal shock, and "linear pyrolysis" in some cases coke residue or pyrolysis residue is formed.

In Ref. [1], materials were divided into coking and non-coking. Such a division is conditional [27], since by introducing certain additives into polymers and changing the conditions of heat and mass transfer, it is possible to convert non-coking material, for example, polymethyl methacrylate, to "coking" during "ceiling" burning.

When classifying coke residues, it is more convenient to use the concepts of crystallinity and amorphism (Figure 6.1. a, b, c, d) models of coke, in which the crystalline phase plays a significant role; e, f, g, h are

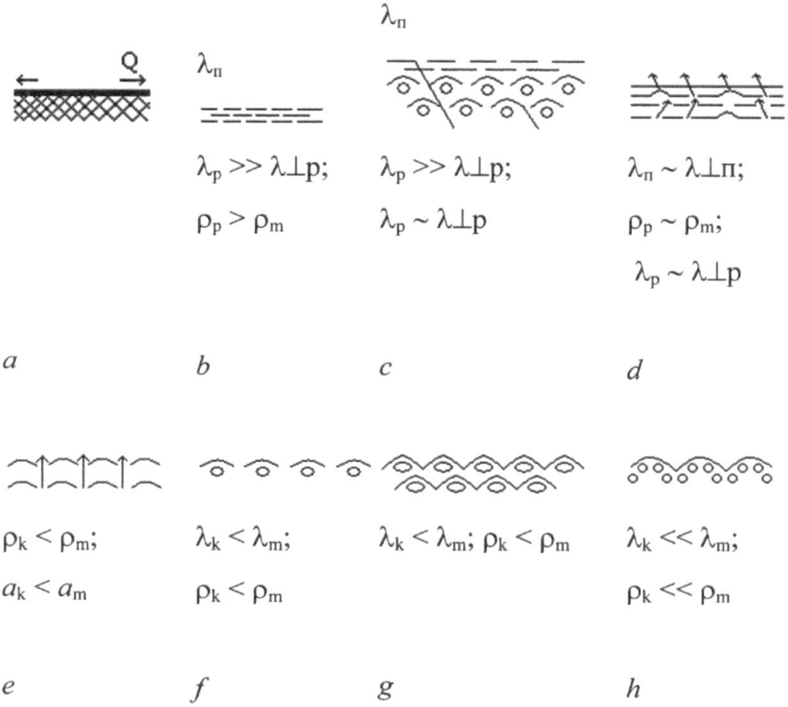

FIGURE 6.1 Coke models (structure of the coke layer): a – coke screen, practically no pores, heat sink over the surface, thermal conductivity over the surface is maximum (idealized case); b – porous coke with a small length of pores, high density, heat flow mainly along the surface; (c) a porous coke of a layered structure consisting of a high-density crust surface with small pores and the maintenance of the condition $\lambda_p >> \lambda \perp p$ and a layer of lower density containing gas bubbles ("coal" foam), which is characterized by "isotropic" of thermal characteristics; (d) porous coke with extended pores, the density and thermal conductivity of coke somewhat differ from the thermal conductivity of the material, part of the pores is filled with amorphous products of material destruction; (e) amorphous porous coke containing a significant percentage of large-diameter pores and craters, coke density less than the density of the source material, the thermal diffusivity of coke also decreases; e – amorphous porous coke containing a significant amount of gas bubbles; g – the same coke of layered structure; h – penoks without pores.

models of amorphous coke, in which amorphous decomposition products). Naturally, during combustion, the thermophysical parameters and the density of coke change. Typically, the coke density with a predominance of the crystalline phase and a small amount of micropores exceeds the density of the source material; the thermal conductivity of this coke is also higher than the thermal conductivity of the source material. It is possible to change the density and thermal conductivity along the normal to the surface, when the heat sink from the source over the coke surface exceeds the thermal diffusivity into the material, and the density of the surface layer is much higher than the density of the subsurface layer. In this case, the change in the parameters ρ, λ, c, a and not occurs in jumps. However, there are cokes of a layered structure consisting of two layers: a layer with a fairly high density, with a small number of micropores of the crust, with high and a and a layer with a lower density and thermal conductivity close to or even lower than the parameters of the source material. In the other three cases, models of amorphous porous cokes or foam coke are given, for which the density and thermal diffusivity are less or significantly less than the corresponding parameters of the starting materials. Shown in Figure 6.1 models of coke in the majority have experimental confirmation.

The authors of work [28] roughly divide the structure of coke formed during the combustion of a number of polymers into two types: dense and weakly porous, and, in their opinion, the type of structure of coke depends mainly on the type of polymer. Thus, the structure of cokes was studied by scanning electron microscopy and showed that less combustible materials (polycarbonate, polyphenylene oxide) during combustion form dense cokes, which provide effective inhibition of heat and mass transfer between the flame and the decomposing polymer.

6.4 FACTORS AFFECTING THE PROCESS OF COKE FORMATION

A number of studies have shown that the formation of coke morphology can be influenced by the melt viscosity, the amount of volatile products, the possibility of crosslinking in a melt of a degrading polymer, the depth of the molten layer [29], and carbonation conditions [30–32].

The processes of coke formation and the morphology of coke are influenced by the addition of various metals, their oxides, and salts. In work [35] it was shown that the most effective additives containing zinc, iron (III), molybdenum (IV), lead to the formation of a carbonized residue (porous on the outside and dense on the inside) when burning polyethylene, polychloroprene, polystyrene, PET. Less effective additives (tin, nickel, barium) lead to the formation of a carbonized residue having pores throughout the volume.

M. Kay and others [33] believe that from a physical point of view, to achieve maximum efficiency, coke should have evenly distributed micropores. A number of authors [34–37] add vitrifying agents such as borates and ammonium phosphates to create a reliable physical barrier in the condensed phase. K. Kishore and K. Mohandes [38] believe that virtually all phosphorus compounds are capable of decomposing into acidic fragments, which then cause the formation of coke of a stable structure. Coke formation in the presence of phosphorus compounds has, in the opinion of the authors, a triple effect: 1) reduction of combustible substances; 2) isolation by coke; 3) creation of a protective layer of non-volatile heat-resistant phosphoric acids, which separate the carbonized layer from oxygen. The formation of a non-volatile carbonized layer that limits the access of oxygen, as well as heat transfer from the flame to the polymer, is one of the approaches to the issue of fire protection [39]. Therefore, you can use physical methods to reduce the flammability of materials based on polymers. In particular, the authors of work [40] obtained high oxygen indices (CI) during heat treatment of epoxy diane compositions without introducing flame retardants into them.

Table 6.3 shows the CI and some mechanical parameters of the investigated epoxy-diane compositions before and after heat treatment. In most cases, heat treatment in the air for several hours at temperatures exceeding the vitrification temperature (Tg) by 20–40°C leads to a two-fold increase in the CI of the material without a noticeable decrease in its mechanical properties.

TABLE 6.3 Change in Physical and Mechanical Characteristics and Flammability of Epoxy Polymers

Item Number	Composition	E, GPa	OI
1	DGER + MFDA (1:1)	2.3/2.45	29.5/56.5
2	ED-20 + MFDA (1:1)	–	27.8/55.1
3	ED-20 + PEPA (8:1)	–	23.0/23.0

Note: In the numerator – the characteristic of the composition before heat treatment, in the denominator – after heat treatment. Heat treatment temperature for composition 1–180°C; 2 and 3–200°C; heat treatment time for the composition of 1–14 h, 2 and 3–9 h.

Currently, in various areas of industry, there is a need for fire protection of metal, wood, and plastic structures and products from them, the creation of foaming coatings is promising in this area [41].

Thermophysical properties of the resulting foam coke determine the effectiveness of flame retardant foaming coatings. In the works [38, 40–44, 47] it was shown that endothermic reactions play a role in reducing the depth of heating of the condensed phase of the coating only with slight foaming, and with a significant increase in volume, they affect only at the initial moment coating work.

Not all polymers are capable of forming coal, coke residue when heated. Polymers prone to carbonization during pyrolysis are characterized by the presence of macromolecules of aromatic carboxylic and/or heterolytic units in the main chain [1]. Carbonization is the process of carbon accumulation in a polymer under the action of various radiations, corrosive media, and heat.

The main parameters affecting the yield of coke during pyrolysis are the chemical nature of the polymer, the heating mode, and the presence of oxygen in the atmosphere [34, 40, 46].

In the case of polymers whose chains consist only of aromatic rings and aliphatic hydrocarbon bridges, the tendency to carbonization increases evenly with an increase in the proportion of aromatic groups in the polymer molecule. In this case, purely aliphatic polymers of coke do not form [47, 48].

The interrelation of the chemical structure of polymers and thermal properties discovered by the authors is interesting [1, 44, 47]. The thermal stability of polymers prone to carbonization and the yield of carbonized residue, the creation of foaming coatings during pyrolysis increase with an increase in the macromolecule of aromatic groups in the main chain.

6.5 FACTORS THAT REDUCE THE PENETRATION OF FOAM COKE

A necessary condition for obtaining a flame retardant foaming coating is the formation of a melt when heated, followed by an irreversible transition to a solid-state.

A stable foam can be obtained if the viscosity of the system increases, so that the gas bubbles do not have time to leave the coating until it hardens, which is expressed by the formula:

$$U_m = U_k K + (1-\beta)\frac{HS}{e}\left(\frac{p_r}{J_r}\Delta p + \frac{p_{zh}}{J_{zh}}\frac{2\sigma}{R}\right)$$

where Um is the mass burning rate; Uk is the rate of coke gasification; K is a Darcy constant; β is the proportion of the polymer converting to coke; H, S is the area and height of the carbonized layer; ρ_g and ρ_l is the density of gaseous and liquid products; ρ is the surface tension; Δp is the difference between internal and external pressure; Jg and Jzh are the viscosity of gases and liquids; R is the outer radius of the through pores.

The movement of liquid degradation products due to capillary forces in the pores penetrating the coke layer, contributes to the development of material combustion, if these products contain a sufficient number of hydrocarbon groups, have low viscosity and low temperatures of the onset of oxidation and decomposition. Therefore, it seems more appropriate to form integral carbon foam on the surface of the protected material, in which the outer layer is dense enough to precede the penetration of gas bubbles and the formation of through pores.

In the work [49] in the course of the experiment, it was found that the presence of phosphorus compounds contributes to the reduction of combustibility, since they contribute to the processes of carbonization during pyrolysis and combustion. In particular, during the pyrolysis of phenol-formaldehyde resin (CFF) with the addition of monoammonium phosphate (MAF), it was found that the presence of a phosphorus compound reduces the Darcy constant several times (Figure 6.3), and this, in turn, reduces the permeability of coke to the products of the lower polymer layers.

The permeability of coke (C), coke obtained by pyrolysis of CFF with the addition of MAF. In Figure 6.2 it can be seen that the permeability of coke from CFF decreases sharply with the introduction of MAF in small amounts of 10–30%. The value of oxygen indices practically does not change. For epoxy diane resin, the value of oxygen indices is higher than that of polystyrene, this is explained by the fact that the products of destruction of epoxy diane resin interact worse with phosphoric acid with the formation of resin substances with low viscosity.

Boron compounds have a similar effect on coke permeability. From Figure 6.3 it can be seen that the permeability of coke decreases under the action of temperatures and reaches its minimum value at a temperature T = 450°C, then rises again, which affects the flammability. It is known that boric acid decomposes when heated to water and boron oxide, whose viscosity at T = 350°C is high, at T = 450°C, the viscosity begins to decrease and covers the pores of coke. Further, with a further increase in temperature, the destruction of the surface films of boron and a sufficiently rapid burn-up of the subsequent layers of coke occurs.

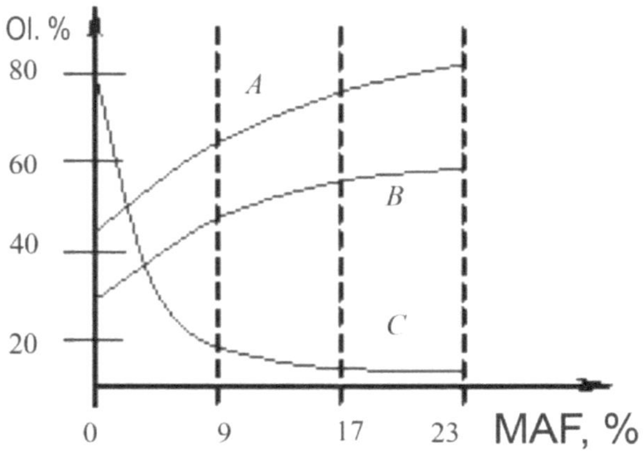

FIGURE 6.2 The combustibility of samples from polystyrene (A) coated with a coke layer of 0.1 cm thick with different permeability (B). Coke obtained by pyrolysis of CFF with the addition of B_2O_3.

Thus, these experiments show that the presence of fluorine-and boron-containing compounds helps to reduce the permeability of the coke protective layer by forming a thin protective film of polyphosphoric acid or boron oxide on its surface. Formed films serve as a barrier for mass transfer of liquid products of polymer destruction through a coke layer to the burning surface.

The greatest physical barrier is given by ammonium borates and phosphates, since they have a vitrifying effect. K. Kishore and K. Mohandes [38] in their work showed that all phosphorus compounds are capable of decomposing into acidic fragments, which cause the formation of coke of a stable structure.

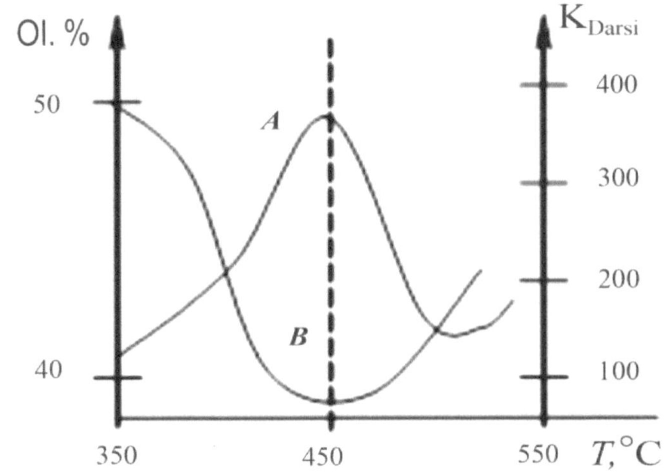

FIGURE 6.3 Dependence of OI and permeability of coke on T.

Also, derivatives of phosphoric acids and carbamide oligomers during the pyrolysis of phenolic-rubber compositions contribute to the formation of network polymeric structures. It is established that the presence of aromatic rings in the composition determines the structure of the formed coke, that is, the aromatic groups are the centers of origin and formation of the structure of coke.

6.6 EVALUATION OF THE EFFECTIVENESS OF INTUMESCENT COATINGS

Evaluating the effectiveness of intumescent fire and heat protective coatings are very important and at the same time very challenging. The resulting carbonized layer changes the conditions of heat and mass transfer between the flame and the non-destructive polymer, which often makes the standard test conditions unattainable. For a more accurate assessment of the effectiveness of intumescent coatings, according to the authors of Refs. [51, 50], the experiment should be carried out either on very small specimens, or by a comparative analysis of the evaluation of the heat-shielding characteristics of the composition with unilateral heating. The use of such a one-dimensional scheme allows, on the one hand, to avoid the influence of heterogeneity of the surface geometry, on the other hand, to evaluate the main characteristic of the protective action of intumescent systems.

The thermophysical and macrokinetic characteristics of physico-chemical transformations determined in a number of works [50, 52] in intumescent material do not allow one to unambiguously describe the physical picture of the processes observed in the experiment due to the incompleteness of knowledge about the mechanism of expansion. This explains the limited number of papers [50, 52–54] devoted to the creation of methods for determining thermal fields and evaluating the effectiveness of fire and heat protective coatings. Usually, such an assessment is carried out under standard heating conditions (standard fire [54]) over the fire resistance limit, defined as the time to reach the specified temperature on the test plate. For steel, this is ≅770 K, and for aluminum, ≅ 520 K [54].

The method for evaluating the effectiveness of coatings described in Refs. [50, 53, 54] reflects the behavior of intumescent materials under standard heating conditions, while various flame retardants are very sensitive to changes in external heat flow. For example, an increase in the rate of heat flow leads to an intensification of the process of expansion (Figure 6.4), therefore, to an increase in the insulating ability of coatings [54].
In this regard, as an assessment of the effectiveness of intumescent polymer coatings, the authors of work [54] proposed the following value:

$$K_{(1)} = \frac{T_m(t) - T_{mp}(t)}{T_m(t) - T_0}$$

The proposed value takes into account the dynamics of temperature change on a thin unprotected metal plate and on a surface protected by an intumescent coating; characterizes the response to external thermal effects. In convective heat transfer, the value can be estimated as follows [54]:

$$T_m(t) - T_0 = (T_\tau - T_0)\left(1 - exp\left(-Bi_m F_{0m}\right)\right) \tag{5}$$

where T_g is the gas temperature in the external flow.

Dependence reflects not only the conditions of external heat exposure, but also the processes in the coating, i.e.,

$$T_{mp} - T_0 = \left(T_\tau - T_0\right) \int \left(\lambda_2 p_2 c_{p2} \frac{\Delta H}{H_0} E_2 Q_2 k_{02} \right) \quad (6)$$

Subtracting (5) from (6) and rationing by the value at each time moment t, it can be assumed that during the comparative tests, the dependence K (t) will characterize only the heat-shielding properties of the intumescent coating.

As an example, in Figure 6.4 shows the results of fire tests of the VPM-2 coating (curve 1) and the SGK-1 coating (curve 2), studied by the authors of [53] (the thickness of the coatings under investigation is 2 mm) in comparison with an unprotected 2.5 mm thick plate (curve 3).

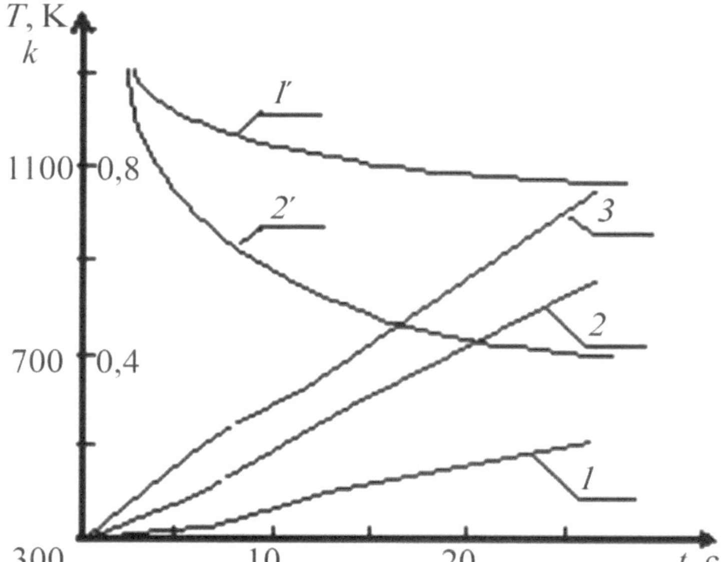

FIGURE 6.4 Temporal dependencies of the temperature of a coated steel plate (curves 1, 2) and uncoated (curve 3), as well as the efficiency coefficient k (curves 1′ and 2′) at a heat flux of 1.8×10^5 W/m^2.

In Figure 6.4, it can be seen that the intumescent SGK-1 coating investigated by the authors [55] is less effective for metal structures than the VPM-2 coating [50]. This shows the change in efficiency ratio (lines 2′ and 1′, respectively).

6.7 CONCLUSION

From the analysis of the literature data, it follows that insufficient attention is paid to the study of the processes occurring during the swelling of composite polymeric materials. Changes in the chemical structure of the coating, the effect on the process of foam and coke formation of certain components, as well as the heating conditions, have not been studied completely. The limited amount of work devoted to these problems is due to the complexity of the processes under consideration, the absence, or the high cost of the necessary technical means.

KEYWORDS

- **fire retardant system**
- **flammability**
- **intumescent coatings**
- **monoammonium phosphate**
- **phenol-formaldehyde resin**
- **vitrification temperature**

REFERENCES

1. Aseeva, P. M., (1981). In: Aseeva, P. M., & Zaikov, G. E., (eds.), *Combustion of Polymeric Materials* (p. 280). M. Science.
2. Kestelman, V. N., (1980). *Physical Methods of Modification of Polymeric Materials* (p. 223). M. Chemistry.
3. Khalturinsky, N. A., (1993). Decrease in flammability of epoxy polymers. *Processing of Polymeric Materials Into Products: Mes.* (pp. 13–14). Report Vseros. Scientific Conf. Izhevsk.
4. Khalturinsky, N. A., (1998). *Physical Methods of Reducing the Flammability of Polymeric Materials, Polymeric Materials of Low Flammability: Mes* (pp. 18–20). Report III International Conf. – Volgograd.
5. Kodolov, V. I., (1976). *Flammability and Fire Resistance of Polymeric Materials: Monograph* (p. 158). M.: Chemistry.
6. Lipanov, A. M., (1992). Experimental modeling of the formation of penoxox. In: Lipanov, A. M., Kodolov, V. I., Larionov, K. I., Shuklin, S. G., & Tyurin, S. A., (eds.). *Materials Symposis: On Burning and Exploding*. Chernogolovka.

7. Kodolov, V. I., Larionov, K. I., Tyurin, S. A., Shuklin, S. G., Dorfman, A. M., Kibenko, V. D., & Lipanov. A. M. (1994). *Pat. RF 2010620 Russian Federation*. The method of preparing the surface of products from organic and fiberglass plastics before applying fire-protective coatings. Publ. 04/15/1994, Byul. No. 7.
8. Glazkova, A. P., (1976). *Catalysis of the Burning of Explosives* (p. 263). M. Science.
9. Nikitina, I. I., (1982). Some features of the carbonization of polymers. In: Nikitina, I. I., Zhubanov, B. A., Gibov, K. M., & Dzhardanova, Z. S., (eds.), *Synthesis of Monomers and Polymers* (pp. 54–77). Alma-Ata: Science.
10. Kodolov, V. I., (1980). *Retardants of Combustion of Polymeric Materials* (p. 272). M. Chemistry.
11. Van Krevelen, B. W., (1977). *Appl. Polym. Sci.: Appl. Polym. Simp., 31*, p. 269.
12. Smirnov, E. P., et al., (1983). *Chem. Physics. 8*, 1109.
13. Breillat, C. J., & Vantelon, J. P. (1983). *Appl. Polym. Sci., 28*(2), 461–471
14. Zhubanov, B. A., (1979). New heat-resistant heterocyclic polymers. In: Zhubanov, B. A., Arkhipova, I. B., & Altabekov, O. A., (eds.), *Alma-Ata: Science* (p. 252).
15. Hastie, J. W., (1973). *J. Res. Nbs., 77A*(6), 733–754.
16. Zhubanov, B. A., (1980). In: Zhubanov, B. A., Altabekov, O. A., & Yu, N., (eds.), *Some Features of Thermal and Thermo-Oxidative Degradation of Polyamides Based on Dicyndrides of Alicyclic Tetracarboxylic Acids with Diamidodiphenyl Oxide* (Vol. 5, pp. 58–62). Izv. AN Kaz. SSR. Ser. "Chemistry."
17. Sultanov, M. T., Sadukov, M. M., & Murashova, U. M., (1986). Inhibition of cellulose burning by phosphorus-containing compounds. *Chemistry, 5*, 35–41.
18. Howard, J., (1978). In: Howard, J., & Sandlers, K., (eds.), *Flame Retardants* (p. 153). Washington: CIEN.
19. Maclaury, M. R., (1979). The influence of platinum fillers and cure on the flammability of peroxide-cured silicone rubber. *J. Fire and Flamm., 10*, 175–198.
20. Edelson, D. (1980). In: Edelson, D., & Kuck, V., (eds.), *Anomalous Behavior of Molybdenum Oxide as a Fire-Retardant for Polyvinyl Chloride* (Vol. 38. pp. 271–293). Comb and Flam.
21. Korablev, G. A., (1999). *The Use of Spatial-Energetic Representations in the Prognostic Evaluation of the Phase Formation of Solid Solutions of Refractory and Related Systems* (p. 290). Izhevsk: IzhGHA.
22. Novikov, S. S., (1968). Modern ideas about the mechanism of condensed systems. In: Novikov, S. S., Pokhil, P. F., & Ryazantsev, Y. S., (eds.), *Firing and Explosion Physics* (pp. 469–482).
23. Lewis, B., Piso, R. N., Taylor, H. S., & Fizmatgiz, L., (1961). *Combustion Processes*, p. 1430.
24. Brauman, S. K., (1979). Char-forming synthetic polymers: Combustion evaluation. *J. Fire Retard. Chem., 6*(4), 244–275.
25. Kishor, K., (1983). Oxygen index and flammability of polymeric materials: A review. In: Kishor. K., & Mohandas, K., (eds.), *J. Sci. and Ind. Res., 42*(2), 76–81.
26. Pokrovsky, S. L., (1985). In: Pokrovsky, S. L., Erin, A. F., Okunev, P. A., & Vasilyev, B. V., (eds.), *The Influence of the Modifying Additive of Furyl Alcohol on the Thermal Stability and Flammability of Carbamide Foams*. Dep. in ONIITE-Chem.
27. Bulgakov, V. K., (1990). In: Bulgakov, V. K., Kodolov, V. I., & Lipanov, A. M., (eds.), *Modeling the Combustion of Polymeric Materials* (p. 238). M. Chemistry.

28. Zhubanov, B. A., (1975). The influence of oxygen concentration on the diffusion burning of polymethyl methacrylate. In: Zhubanov, B. A., Dovlichin, T. K., & Gibov, K. M., (eds.), *Naval Science*, *175*(10), 746–748.
29. Chen, X., (1999). The growth patterns and morphologies of carbon micro-coils produced by chemical vapor deposition. In: Chen, X., & Motojima, S., (eds.), *Carbon, 37*(11), 1817–1823.
30. Zhubanov, B. A., (1975). *Influence of Oxygen Concentration on the Diffusion Burning of Polymethyl Methacrylate, B.A., 175*(10), 746–748.
31. Cherednik, E. M., (1983). The influence of the carbonization conditions of the pitch on some properties of their cokes. In: Cherednik, E. M., Butrin, G. M., & Zimina, L. A., (eds.), *Chemistry of Solid Fuel*, *1*, 74–81.
32. Stueta, D. E., (1975). Polymer combustion. In: Stueta, D. E., Diedwargdo, A. H., Zitomer, F., & Barnes, B. P., (eds.), *J. Polym. Soi., Polym. Chem. Ed.*, *3*(3), 585–621.
33. Kay, M. A., (1979). Review of intumescing materials, with emphasis on melamine formulations. In: Kay, M., Frice, A. F., & Lavery, J., (eds.), *J. Fire Retard. Chem.*, *6*, 69–91.
34. Brauman, S. K., (1980). Char-forming synthetic polymers modification by halogen introduction and use of dehydrohalogenation agents to promote charring. *J. Fire Retard. Chem., 7*(3), 119–129.
35. Brauman, S. K., (1979). Char-forming synthetic polymers, combustion evaluation. *J. Fire Retard. Chem., 6*(4), 244–275.
36. Brauman, S. K., (1977). Polymer degradation and combustion. *J. Polim. Chem., 15*(6), 1507–1509.
37. Mamsumoto, T., (1968). Nonsteady thermal decomposition of plastics. In: Takenori, M., Tochitaka, E., & Jiro, K., (eds.), *12th Sympos. (Internat.) Combust Poitiers* (pp. 93–95). Pittsburgh. Pa. Combust. Inst.
38. Kishore, K., (1982). In: Kishore, K., & Mohandes, K., (eds.), *Action of Phosphorus on Fire Retardancy of Cellulosic Materials, 6*(2), 54–58.
39. Read, R. T., (1985). Mechanisms of flame retardancy. *Polymers Point Color Journal, 175*(4), 213–214.
40. Mark, H. F., (1975). Combustion of polymers and retardation. In: Mark, H. F., Atlas, S. M., Shalaby, S. W., & Pearse, E. M., (eds.), *Polim. News, 2*(5/6), 3–12.
41. Nechvolodov, E. M., (1987). The effect of heat treatment on the combustion process of epoxyamine polymers. In: Nechvolodov, E. M., Galchenko, A. G., & Rogovina, S. Z., (eds.), *Chem. Physics*, (5/6), 696.
42. Glebov, K. M., (1977). Fire retardant compositions based on epoxy resin. In: Glebov, K. M., Kapyrina, V. Y., & Dovlichin, T. K., (eds.), *Plastics*, *12*, 46–47.
43. Dovlichin, T. K., (1979). Fireproof polymer coatings. In: Dovlichin, T. K., Zhubanov, B. A., & Gibov, K. M., (eds.), *Chemistry and Physical Chemistry of Polymers-Alma-Ata: Science*, 43–56.
44. Dovlichin, T. K., (1979). The thermal regime of foaming fire-retardant coatings. In: Dovlichin, T. K., Zhubanov, B. A., Nikitina, I. I., Mamleev, V. S., & Gibov, K. M., (eds.), *Chemistry, 5*, 36–40.
45. Kipling, J. S., (1966). Factors affecting the graphitization of car-bon: Evidence from polarized light microekopy. In: Kipling, J. S., & Shooter, P. V., (eds.), *Carbon, 4*(1), 1–4.

46. Kulev, D. K. (1985). Polymer materials with low flammability and smoke-forming ability. *Plastics, 10*, 51–52.
47. Mochida, I., (1980). Carbonization Reactions of Organic Compounds. Mochida, I., Yozo, K., & Kenjiro, T., (eds.). DOI: 10.5059/yukigoseikyokaishi, *38*(5), 433–446.
48. Fabris, H. J., (1977). Flammability of elastomeric materials. In: Fabris, H. J., & Som-mer, J. C., (eds.), *Rubber Chem. and Technol., 50*(3), 523–569.
49. Perrins, L. E., (1974). Measurement of flame spread velocities. In: Perrins, L. E., & Pettett, K. J., (eds.), *Fire and Flammability, 5*(1), 85.
50. Burning Behavior Plastics, During and After Contact with Glowing Rod. Div 53459 Deutsches Institut fur Normung, German institutes for rationing.
51. Khalturinsky, N. A., (1984). Polymer burning and the mechanism of action of flame retardants. In: Khalturinsky, N. A., Popova, T. V., & Berlin, A. A., (eds.), *Chemistry Advances, 33*(2), 326–346.
52. Hallman, J. R., (1972). Ignition of polymers society of plastics engineers. In: Hall-man, J. R., Welker, J. R., & Sliepcepcevich, C. M., (eds.), *30th Annual Technical Conference* (p. 283). Palmer House, Chicago.
53. Pohil, P. F., (1953). On the mechanism of combustion of smokeless powders. *Explosion Physics, 2*, 181–212.
54. Lewis, B., Piso, R. N., Taylor. H. S., & Fizmatgiz, L., (1961). *Combustion Processes*, p. 430.
55. Williams, F. A., (1976). Mechanism of fire spread, 16th Symposium (Int.) on combustion. *Combust. Jnst.*, p. 1281.

CHAPTER 7

OPTIMIZATION OF WEAR RATE ON THE LOW-COST REINFORCED-HYBRID ALUMINUM METAL MATRIX COMPOSITE

L. FRANCIS XAVIER, G. RAVICHANDRAN, and N. SANTHOSH

Assistant Professor, Department of Mechanical and Automobile Engineering, Christ (Deemed-to-be-University), Kanminike, Kumbalgodu, Mysore Road, Kengeri, Bengaluru – 560074, Karnataka, India, E-mails: francis.xavier@chirstuniversity.in (L. Francis Xavier), ravichandran.g@chirstuniversity.in (G. Ravichandran), santhosh.n@chirstuniversity.in (N. Santhosh)

ABSTRACT

As there is an increasing demand for low-cost reinforcement particles in preparing composite materials, in this work discarded waste black toner powder and coconut shell ash particles were used as the reinforcement materials to prepare hybrid aluminum metal matrix composite (AMMC). As wear is one of the utmost universally faced industrial delinquent, wear test was performed on the prepared composites. Further, Taguchi based design of experiments and analysis of variance (ANOVA) was used to discern the most appropriate levels of the wear parameters. Regression analysis was also used to recognize the link between the wear rate and the parameters taken for the investigation. From the experimental outcomes, applied load and sliding distance was recognized as the most important parameters influencing the wear rate.

7.1 INTRODUCTION

In this modern world, aluminum metal matrix composite (AMMC) are extensively used in aerospace, defense, sports, and in many industrial applications for many of its splendid properties like high strength, thermal conductivity stiffness and combined properties like wear resistance with fracture toughness and high strength with good corrosion resistance [1].

Generally, Sic, Al_2O_3, TiB_2, B4C are the commonly used reinforcement materials in preparing the AMMC. Among the different types of reinforcement, a particulate form of ceramics reinforced with AMMC have attractive and desirable properties like the ease of fabrication, oxidation resistance, capable of withstanding higher operating temperature when compared to the other geometries of reinforcement such as fiber and flakes [2]. Currently, the preparation of AMMC at low cost is gaining more attention among materials researchers around the globe. From the Literature, to mention a few, Rice husk ash [3], Breadfruit seed hull ash [4], Biomass waste materials like palm oil clinkers [5], Fly ash [6], Wet Grinder stone dust particle [35] and Tonner Powder [34] are some of the works related with utilization of low-cost reinforcement materials in preparation of composites.

The application of AMMC is restricted due to its poor resistance to wear rate under dry lubrication conditions [7, 8]. Wear is considered as one of the universally faced industrial problems where the material is affected mainly by the applied load, speed, and environmental conditions [9]. Wear is a slow and progressive volume loss of material from a solid surface which is subjected to repetitive rubbing action [10, 34]. The major concern with wear is that it causes a huge quantity of expenditure caused due to the replacement or servicing the worn-out parts or equipment [11]. According to Ref. [12], the wear resistance of a metal matrix composite is influenced by various microstructural characteristics like particle size, shape, quantity fraction, and distribution of strengthening material in the base matrix.

In this work, waste black tonner and coconut shell dust particles were used as reinforcement materials in the preparation of AMMC. Wear studies were conducted on the prepared AMMC. Taguchi based Optimization method and analysis of variance (ANOVA) was used to minimize the wear rate.

The experiments were planned and designed based on Taguchi's based optimization method. Taguchi-based design of experiments is considered as a powerful experimental design technique which can be effectively used to reduce the number of trails/repeating of the experiments, cost, and valuable time [15]. Taguchi method is a very effective method when compared to the time consuming and complicated traditional experimental design methods [16]. Taguchi-based optimization method has been effectively used by several researchers to study the wear characteristics of AMMC [17–21]. In this investigation, the Taguchi method was applied to decide the optimal level of the parameters for obtaining a minimal wear rate. Further, the percentage impact of specific parameters taken for the wear test and their influence on the wear rate was determined using ANOVA [18].

7.2 MATERIALS AND METHODS

In this present investigation, Al6063 alloy was selected as the base matrix material for preparing the composite. Al6063 alloys are typically used in aircraft applications, extrusions, architectural applications, window frames, and irrigation tubing [13]. Waste black toner particles and coconut shell ash particles of 400 mesh size were used as reinforcement particles. From the chemical analysis of the waste black toner powder, the elements like: Si, C, and Fe were found to be its major ingredients. Whereas SiO_2, Al_2O_3, $Fe_2O_3$and MgO were observed as the major ingredients present in the coconut shell ash.

Three samples were taken for the investigation. Sample-I = Al6063 alloy, Sample-II = Al6063 alloy + 5 wt.% of waste black toner particles and 5 wt.% of coconut shell ash particles; Sample-III = Al6063 alloy + 10 wt.% of waste black toner particles and 10 wt.% of coconut shell ash particles. The composites were prepared by stir casting method.

Pin-on-disc test apparatus (Model TR 20-LE, Ducom) was used for conducting the wear test. The specimens were machined to pin size of 10 × 10 × 25 mm. The wear test was piloted at standard room temperature (RT) with the load ranging from 9.81, 19.62, and 29.43 N at a sliding speed of 1.57, 3.14 and 4.71 m/s and with a sliding distance of 1000, 2000, and 3000 m (Figure 7.1).

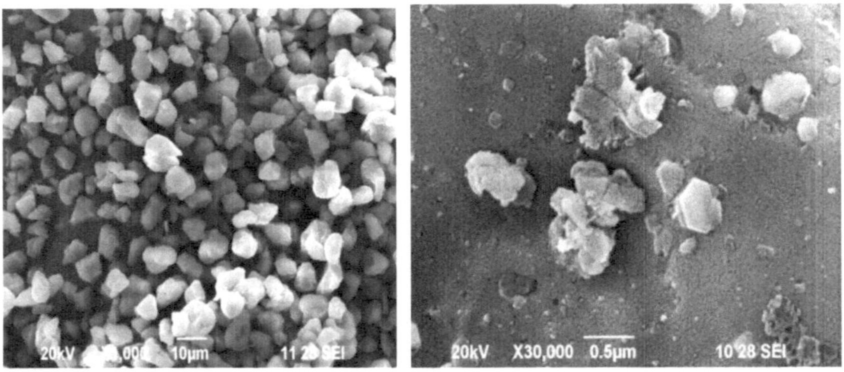

FIGURE 7.1 (a) SEM image of waste tonner particle; (b) SEM image of coconut shell ash particles.

7.3 PLAN OF EXPERIMENTS

Four control factors namely: (1) sliding speed (m/s) (2) load (N) (3) sliding distance (m), and (4) working material (%) were taken for the investigation. All the four factors at three levels and the interactions between load × working material, sliding speed × working material and sliding speed × load are to be studied.

In this work, the requisite degree of freedom for studying the key and interaction effect is 26. As the degrees of freedom for the orthogonal array is larger than the number of degrees of freedom necessary for swotting the key and interaction effect, L27 orthogonal array was selected [19]. The levels and the values of the wear parameters used for conducting the wear test were determined by conducting preliminary experiments. Table 7.1 indicates the control factors for wear rate with their levels.

TABLE 7.1 Control Factors Taken for Optimizing the Wear Rate and Their Levels

Symbol	Control Factors	Levels		
		I	II	III
S	Sliding Speed (m/s)	1.57	3.14	4.71
L	Load (N)	9.81	19.62	29.43
D	Sliding Distance (m)	1000	2000	3000
R	Working Material (%)	1	2	3

7.4 RESULTS AND DISCUSSIONS

Table 7.2 illustrates the investigational results which were obtained by conducting the experiments as per the L27 orthogonal array. The experiments were planned with an objective to relate the effect of load, working material, sliding speed, and sliding distance on the wear rate. The experimental results were analyzed using MINITAB 16 software. The wear rate is the response to be investigated with an objective as smaller-the-better.

Figures 7.2(a), 7.3(a) and 7.4(a) indicate the SEM image of the tattered surface of all the samples taken for the investigation. From the SEM images, as shown in Figures 7.3(a) and 7.4(a), it is clear that the tattered surface of the composites are covered with near-continuous oxide films when compared to the Al6063 alloy, coincides with the earlier findings on Al/SiCp composites [22]. From Figure 7.3(a), we can see the formation of a mechanically mixed layer (MML) with a higher amount of oxide content spread more uniformly distributed surface of sample-II.

The MML plays a vital part in aggregating the resistance to the wear rate. The MML may be formed due to the relocation and mixing of reinforced ingredients when conducting the wear test under a certain velocity range and load [24]. The MML turns as a protective shield, prevents the metal-to-metal interaction amongst the sliding surfaces, and thus reduces the wear rate and friction coefficient of AMMC [25, 26]. Further, the presence of Fe and hard alumina particles in sample-II known for its good thermal stability [28] plays a vital role in increasing the load-bearing capacity and has retained its strength, resistance to thermal softening at elevated temperature. Thus, the sample-II has shown good resistance to adhesive wear compared with the other samples taken for the investigation. According to Ref. [27], the Fe rich film has a low coefficient of friction, contributes to better resistance to wear rate of the composite and coincides with our findings. The energy-dispersive x-ray spectroscopy (EDAX) test report of the tattered surface of sample-II, as shown in Figure 7.3(b) confirms strong Fe and oxygen peaks when compared with Figures 7.4(b) and 7.2(b).

TABLE 7.2 Orthogonal Array of Taguchi for Wear Rate

Test	Sliding Speed, S (m/s)	Load, L (N)	Sliding Distance, D (m)	Working Material, R (%)	Wear Rate (mm)
1.	1.57	9.81	1000	1	0.02498
2.	1.57	9.81	2000	2	0.02485
3.	1.57	9.81	3000	3	0.03098
4.	1.57	19.62	1000	2	0.02754
5.	1.57	19.62	2000	3	0.02897
6.	1.57	19.62	3000	1	0.03124
7.	1.57	29.43	1000	2	0.03114
8.	1.57	29.43	2000	1	0.03245
9.	1.57	29.43	3000	2	0.03415
10.	3.14	9.81	1000	2	0.02321
11.	3.14	9.81	2000	3	0.02401
12.	3.14	9.81	3000	1	0.03012
13.	3.14	19.62	1000	3	0.02641
14.	3.14	19.62	2000	1	0.02857
15.	3.14	19.62	3000	2	0.03112
16.	3.14	29.43	1000	1	0.02854
17.	3.14	29.43	2000	2	0.02754
18.	3.14	29.43	3000	3	0.03324
19.	4.71	9.81	1000	3	0.02301
20.	4.71	9.81	2000	1	0.02368
21.	4.71	9.81	3000	2	0.02841
22.	4.71	19.62	1000	1	0.02415
23.	4.71	19.62	2000	2	0.02845
24.	4.71	19.62	3000	3	0.03018
25.	4.71	29.43	1000	2	0.02714
26.	4.71	29.43	2000	3	0.02645
27.	4.71	29.43	3000	1	0.03125

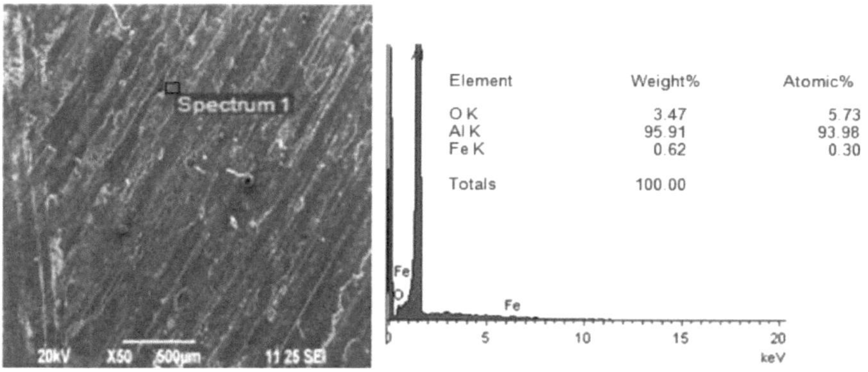

FIGURE 7.2 (a) SEM image of sample-I, at an applied load of 29.43 N, sliding speed of 4.71 m/s after running 1000 m (b) EDAX test report at the region marked as spectrum 1on the sample-I.

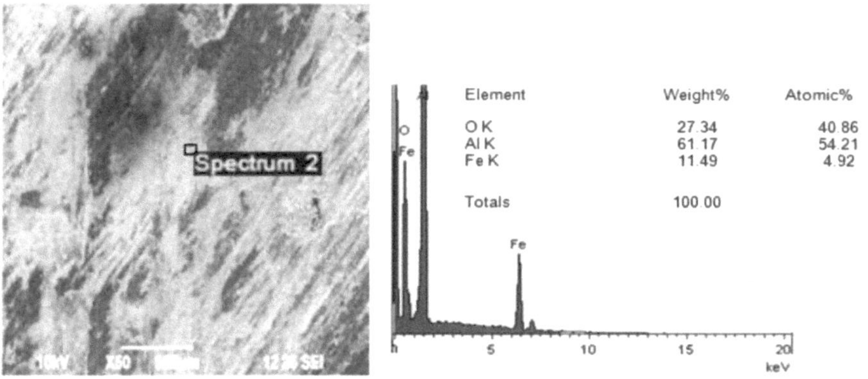

FIGURE 7.3 (a) SEM image of sample-II, at an applied load of 29.43 N, sliding speed of 4.71 m/s after running 2000 m (b) EDAX test report at the region marked as spectrum 2 on the sample-II.

Among the three samples taken for the investigation, the hardness of the Sample-I is less when compared with the Sample-II and Sample-III. According to Ref. [29], adhesive wear is inversely proportionate to the hardness of the material and directly proportional to the applied load and sliding distance. In the case of sample-I due to the lack of reasonable strength and hardness, the matrix gets easily softened at higher loads and sliding speed, prevents the formation of stable MML layer and thus increases the wear rate drastically.

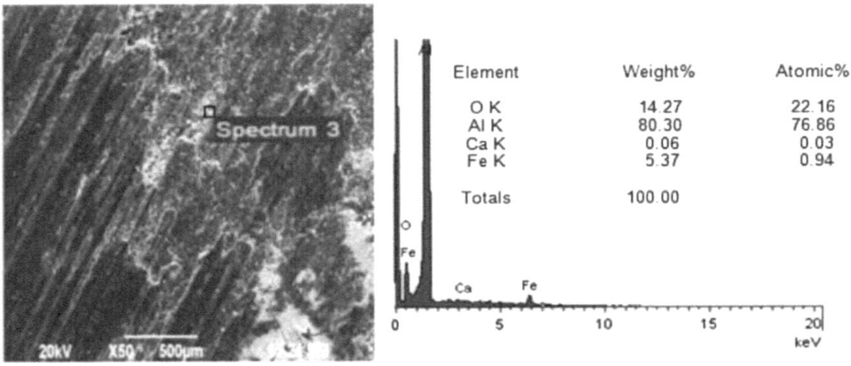

FIGURE 7.4 (a) SEM image of sample-III, at an applied load of 29.43 N, sliding speed of 4.71 m/s after running 3 000 m (b) EDAX test report at the region marked as spectrum 3 on the sample-III.

On examining the tattered surface of sample-I, as shown in Figure 7.2(a), it is clear that the formation of oxide films as discontinuous and can be seen only in certain regions. Deep grooves are found to be more severe on the sample-I when compared with the other samples. The deep grooves may be formed due to the penetration of the hard asperities onto the soft pin surface. Further, when conducting the wear test by increasing the sliding distance from 1000 m to 3000 m the wear rate would increase drastically. Thus the wear rate of the Sample-I intensifies due to increased metal-to-metal interaction owing to the absence of protective oxide film. The EDAX test report of the wear tracks of sample-I as shown in Figure 7.2(b) confirms small Fe and oxygen concentration.

Figure 7.4(a) reveals the SEM image of the tattered surface of sample-III, showing the oxide layer formation on the sample-III as thicker and discontinuous. From the SEM image, we can see shallow grooves running parallel to the sliding direction filled with broken oxide layers, flaky debris, and pores giving an indication of abrasive wear. According to Ref. [30], the pore can impart certain beneficial results on the wear rate of the samples by trapping the wear debris inside the pores, reduce the possibility of particle agglomeration and also prevents the metal-to-metal contact [30, 31]. Thus, the wear rate of sample-III is comparatively less compared with the unreinforced sample-I even after running a longer distance.

7.5 ANALYSIS OF VARIANCE (ANOVA)

In this work, ANOVA was used to find the most suitable combination of the control factors for minimal wear rate by analyzing the relative significance in terms of their percentage of impact (p) on the wear rate. Further, the interactions between the sliding speed × load; sliding speed × working material; load × working material on the response were also studied. The executed investigational plan was analyzed at a confidence level of 95% [20]. Table 7.3 reveals the results of ANOVA for wear rate. The attained R^2 value for wear rate was 99.0, which is desirable [21]. From Table 7.3, we can comprehend that the p-value for sliding speed, load, sliding distance, the interaction between sliding speed × working material is less than or equal to 0.05 which shows that the factors are important and have a substantial impact on the response [32]. The working material and the interaction among sliding speed × load and load × working material have a p-value greater than 0.05 which shows that they are insignificant and do not have any substantial impact on the response i.e., wear rate.

TABLE 7.3 ANOVA Results for Wear Rate

Source	DOF	Seq SS	Adj SS	Adj MS	F	P	Percent-Age of Contribution
Sliding speed (m/s)	2	2.9177	2.9177	1.45885	32.38	0.001	11.26
Load (N)	2	8.3912	8.3912	4.19558	93.13	0.000	32.39
Sliding distance (m)	2	11.7313	11.7313	5.86564	130.20	0.000	45.28
Working material (%)	2	0.0325	0.0325	0.01624	0.36	0.711	0.125
Sliding speed (m/s) × Load (N)	4	0.5288	0.5288	0.13220	2.93	0.116	2.04
Sliding speed (m/s) × Working material (%)	4	1.5902	1.5902	0.39756	8.82	0.011	6.14
Load (N) × Working material (%)	4	0.4470	0.4470	0.11175	2.48	0.154	1.73
Residual Error	6	0.2703	0.2703	0.04505			1.04
Total	26	25.9089					100

It is clear from Table 7.3 that sliding distance (D) is the most significant parameter increasing the wear rate and contributes (p=45.28%) on the wear rate followed by the load (p = 32.39%) and sliding speed (p = 11.26%). The working material contributes only (p = 0.125%) which shows that there is no significant increase in the wear rate by increasing the reinforcement material. The interaction between sliding speed × working material contributes (p = 6.14%) and sliding speed × load contributes (p = 2.04%). Further, the interaction between load × working material contributes only (p = 1.73%). The error associated with the ANOVA table was approximately about 1.04%.

7.6 REGRESSION ANALYSIS

Regression analysis is an arithmetic tool, which can be efficiently used to identify the liaison among the variables selected for investigation [33]. Equation (1) was developed using the MINITAB 16 software.

$$Wear = 0.0216 - 0.000834S + 0.000219L + 0.000002D - 0.000003R \quad (1)$$

where; S = Sliding Speed (m/s), L = Load (N), D = Sliding distance (m), R= Working material (%).

In this work, the R^2 value of the regression coefficient is 0.84 which indicates that the model as fitted and explains 84.0% of the variability for wear rate. From Eq. (1), the negative value of the coefficient indicates that the wear rate drops down with an increase in the associated parameters whereas the positive value of the coefficients gives an indication that the wear rate increases with an increase in the associated parameters [18, 34].

By substituting the recorded values of the parameters in Eq. (1), the wear rate of the samples can be calculated. Further, the weightage of each parameter on the wear rate are indicated by their magnitude. It is clear from Eq. (7.1) that the sliding speed as the most significant parameter having more influence on the wear rate followed by load, working material and sliding distance. The coefficient associated with sliding speed and working material is negative, which indicates that the wear rate drops down with an increase in the sliding speed and working material content. The upsurge in the wear rate with the rise in sliding speed may be due to the rise in the interfacial temperature caused due to the frictional heating at the contact surfaces, leading to a higher rate of oxidation, softens the surface of the composite, and encourages adhesion between the sliding disc and the pin. Whereas the coefficient related with sliding distance and Load is positive, which shows that the wear rate increases with an increase in the sliding distance and applied Load.

7.7 CONFIRMATION TEST

The confirmation test is the crucial step to validate the experimental results and it is immensely recommended by Taguchi [15, 34]. In this study, the confirmation test was executed by picking the set of factors as presented in Table 7.4. The confirmation test was conducted and the results were compared with the regression Eq. (7.1). From the confirmation test as shown in Table 7.5, we can see that the errors associated with wear rate range between 4.43% and 8.61%. Therefore, we can conclude that the developed regression equation for wear rate correlates with the experimental values with a reasonable degree of approximation.

TABLE 7.4 Values Used in the Confirmation Test for Wear Rate

Test	Sliding Speed (m/s)	Load (N)	Sliding Distance (m)	Working Material (%)
1	2.25	14.5	700	1
2	3.14	24.5	1200	2
3	4.25	29.5	2350	3

TABLE 7.5 Result of the Confirmation Test for Wear Rate (mm)

Test	Experimental Value Wear Rate (mm)	Value Predicted Using the Regression Eqn. (7.1) for Wear Rate (mm)	Percentage of Error (%)
1	0.02658	0.02429	8.61%
2	0.02798	0.02674	4.43%
3	0.02654	0.02451	7.64%

7.8 CONCLUSIONS

In this work, low cost reinforced AMMC with improved wear resistance which is suitable for automobile applications was prepared by reinforcing discarded waste black tonner dust particles and coconut shell ash particles. From the wear test, the following observations can be made.

- A very nominal variation was observed between the experimental and the predicted value, which indicates that the Taguchi method and ANOVA have been successfully used to pinpoint the optimal levels of wear parameters to obtain a minimal wear rate.
- From the wear test results, the sliding distance, and the applied load was identified as the most significant parameters influencing the wear rate.
- Among the prepared samples taken for the investigation, the sample-II =Al6063 alloy + 5 wt.% of waste black toner particles and 5 wt.% of coconut shell ash particles has shown good resistance to wear rate even at higher loads and sliding speed.
- The wear rate of both the unreinforced Al6063alloy and the prepared composites increases with the increase in applied load and decreases with an increase in the sliding speed.

KEYWORDS

- **analysis of variance**
- **coconut shell ash**
- **dry sliding wear test**
- **regression analysis**
- **Taguchi method**
- **waste black tonner**

REFERENCES

1. Basavarajappa, S., & Chandramohan G., (2005). Dry sliding wear behavior of hybrid metal matrix composites. *Materials Science, 11*(3), 253–257.
2. Kaw, A. K., (2006). *Mechanics of Composite Materials [M]*. New York: Taylor & Francis Group.
3. Alaneme, K. K., & Olubambi, P. A., (2013). Corrosion and wear behavior of rice husk ash-alumina reinforced Al-Mg-Si alloy matrix hybrid composites. *Journal of Materials Research and Technology, 2*(2), 188–194.

4. Atuanya, C. U., Ibhadode, A. O. A., & Dagwa, I. M., (2012). Effect of breadfruit seed hull ash on the microstructures and properties of Al-Si-Fe alloy/breadfruit seed hull ash particulate composites. *Results in Physics, 2*, 142–149.
5. Zamri, Y. B., Shamsul, J. B., & Amin, M. M., (2011). Potential of palm oil clinkers as reinforcement in aluminum matrix composites for tribological applications. *International Journal of Mechanical and Materials Engineering, 6*, 10–17.
6. Rajan, T. P. D., Pillai, R. M., Pai, B. C., Satyanarayana, K. G., & Rohatgi, P. K., (2007). Fabrication and characterization of Al-7Si-0.35 Mg/fly ash metal matrix composites processed by different stir casting routes. *Composites Science and Technology, 67*(15), 3369–3377.
7. Habibolahzadeh, A., Hassani, A., Bagherpour, E., & Taheri, M., (2013). Dry friction and wear behavior of in-situ Al/Al3Ti composite. *Journal of Composite Materials, 48*(9), 1049–1059.
8. Veeresh, K. G. B., Rao, C. S. P., & Selvaraj, N., (2012). Studies on mechanical and dry sliding wear of AL6061-SiC composites. *Composites: Part, B., 43*(3), 1185–1191.
9. Ferit, F., Sakip, K., Ramazan, K., & Omer, S., (2011). Investigation of unlubricated sliding wear behaviors of in-situ ALB2/AL metal matrix composite. *Advanced Composite Letters, 20*(4), 109–116.
10. Lancaster, L., Lung, M. H., & Sujan, D., (2013). Utilization of agro-industrial waste in metal matrix composites: Towards sustainability. *International Journal of Environmental, Ecological, Geological and Mining Engineering, 7*(1), 25–33.
11. Mukundadas, P. K., Kanakuppi, S., Gundenahalli, P. P., & Satyappa, B., (2006). Dry sliding wear behavior of garnet particles reinforced zinc-aluminum alloy metal matrix composites. *Materials Science, 12*(3), 209–213.
12. Ravi, K. K., Mohanasundaram, K. M., Arumaikkannu, G., & Subramanian, R., (2012). Analysis of parameters influencing wear and frictional behavior of aluminum-fly ash composites. *Tribol. Trans., 55*, 723–729.
13. Hemalatha, K., Venkatachalapathy, V. S. K., & Alagumurthy, N., (2013). Processing and synthesis of metal matrix Al 6063/Al_2O_3 metal matrix composite by stir casting process. *Int. J. of Engineering Research and Applications, 3*(6), 1390–1394.
14. Mishra, S. B., Chandra, K., & Prakash, S., (2013). Dry sliding wear behavior of nickel-iron and cobalt-based super alloys. *Tribology, 7*(3), 122–128.
15. Ross, P. J., (1996). *Taguchi Techniques for Quality Engineering*. New York: McGraw-Hill Book Company.
16. Satapathy, A., & Patnaik, A., (2008). Analysis of dry sliding wear behavior of red mud filled polyester composite using Taguchi method. *Journal of Reinforced Plastics and Composite, 29*(24), 2883–2897.
17. Sahin, Y., (2005). The prediction of wear resistance model for the metal matrix composites. *Wear, 258*(11/12), 1717–1722.
18. Basavarajappa, S., & Chandramohan, G., (2005). Wear studies of metal matrix composites: A Taguchi approach. *J. Mater. Sci. Technol., 21*(6), 845–850.
19. Satpalkundu, R. B. K., & Ashok, K. M., (2013). Study of dry sliding wear behavior of aluminum/Sic/Al_2O_3/graphite hybrid metal matrix composite using Taguchi technique. *International Journal of Scientific and Research Publications, 3*(8), 1–8.

20. Bhaskar, H. B., & Abdul, S., (2012). Dry sliding wear behavior of aluminum/ $Be_3Al_2(SiO_2)6$ composite using Taguchi Method. *Journal of Minerals and Materials Characterization and Engineering, 11*, 679–684.
21. Aman, A., Hari, S., Pradeep, K., & Manmohan, S., (2008). Optimizing power consumption for CNC turned parts using response surface methodology and Taguchi's technique - A comparative analysis. *Journal of Materials Processing Technology, 200*(1/3), 373–384.
22. Wilson, S., & Alpas, A. T., (1997). Wear mechanism maps for metal matrix composites. *Wear, 212*(1), 41–49.
23. Shanthi, M., Nguyen, Q. B., & Gupta, M., (2010). Sliding wear behavior of calcium containing AZ1B/Al_2O_3 nano composites. *Wear, 269*(5), 473–479.
24. Venkataraman, B., & Sundararajan, G., (2000). Correlation between the characteristics of the mechanically mixed layer and wear behavior of aluminum, Al7075 alloy and Al-MMCs. *Wear, 245*(1/2), 22–38.
25. Straffelini, G., Pellizzari, M., & Molinari, A., (2004). Influence of load and temperature on the dry sliding behavior of Al-based metal matrix composites. *Wear, 256*, 754–763.
26. Venkataraman, B., & Sundararajan, G., (2000). Correlation between the characteristics of the mechanically mixed layer and wear behavior of aluminum, Al-7075 alloy and Al-MMCs. *Wear, 245*, 22–38.
27. Zhang, J., & Alpas, A. T., (1993). Wear regimes and transactions in Al_2O_3, particulate-reinforced aluminum alloys. *Mater. Sci. Eng. A, 161*(2), 273–284.
28. Hassan, S. F., Tan, M. J., & Gupta, M., (2008). High-temperature tensile properties of Mg/Al_2O_3 nano composite. *Mater. Sci. Eng. A, 486*(1), 56–62.
29. Hassan, S. F., Tan, M. J., & Gupta, M., (2011). High-temperature tensile properties of Mg/Al_2O_3 nano composite. *American Journal of Scientific and Industrial Research, 2*, 99–106.
30. Lin, R. Y., & Deshpande, P. K., (2006). Wear resistance of WC particle reinforced copper matrix composites and the effect of porosity. *Journal of Materials Science and Engineering, A., 418*(1/2), 137–145.
31. Stott, F. H., & Wood, G. C., (1978). The influence of oxides on the friction and wear of alloys *Tribol. Int., 11*, 211–218.
32. Rajesh, S., Rajakarunakaran, S., & Sudhakara, P. R., (2012). Modeling and optimization of sliding specific wear and coefficient of friction of aluminum based red mud metal matrix composite using Taguchi method and Response surface methodology. *Materials Physics and Mechanics, 15*, 150–166.
33. Radhika, N., Vaishnavi, A., & Chandran, G. K., (2014). Optimization of dry sliding wear process parameters for aluminum hybrid metal matrix composites. *Tribology in Industry, 36*, 188–194.
34. Francis, X. L., & Suresh, P., (2016). Studies on dry sliding wear behavior of aluminum metal matrix composite prepared from discarded waste particles. *International Journal of Advanced Engineering Technology, VII*(I), 539–543.
35. Francis, X. L., & Suresh, P., (2016). Wear behavior of aluminum metal matrix composite prepared from industrial waste. *The Scientific World Journal*, 1–8.

CHAPTER 8

ELASTIC-PLASTIC AND CREEP TRANSITION IN STRUCTURAL COMPONENTS

SHIVDEV SHAHI,[1] PANKAJ THAKUR,[2] VANDANA GUPTA,[3] and SATYA BIR SINGH[1]

[1]*Department of Mathematics, Punjabi University, Patiala – 147002, India*

[2]*Department of Mathematics, ICFAI University, Solan, Himachal Pradesh, India*

[3]*Department of Mathematics, Dashmesh Khalsa College, Zirakpur (Mohali), Punjab, India*

ABSTRACT

Elastic-plastic and creep transition in structural components have been studied. In elastic solids, the state of strain depends only on the final state of stresses. On the other hand, the deformation in plastic solids is determined by the complete history of loading in which the distortion in the material is obtained as the sum of incremental distortions following the strain path. Material, when constrained to a load for a very long time at moderately elevated temperatures, show progressive deformation is known as creep. The non-linear character was studied by Seth considering the transition surface function and generalized strain measure to determine elastic-plastic and creep state. A number of elastic-plastic and creep problems pertaining to various structural components, made of materials exhibiting different kinds of isotropy and anisotropy have been solved using this transition theory. This chapter reviews development in the mechanical behavior of structural components using both types of theories, i.e., classical as well as transition theory. It is sufficient to say that the transition functions which define the non-linearity of elastic-plastic and creep transition are more accurate when compared to the classical theory.

8.1 HISTORICAL PERSPECTIVES

Mechanics of solids is a basic discipline, which is as yet poorly understood in respect of phenomena in mechanical response and failure of materials and structures. The research in solid mechanics is essential not only for a basic understanding of mechanical phenomenon but also for advancing engineering methodology and structural technology. Advances in the subject are central to assure safety, reliability, and economy in the design of devices, structures, and complete systems that are essential to the continued development of power generation technology such as fusion, nuclear, and gas turbine power, aerospace, and surface transportation vehicles, earthquake resistant design, offshore structures, orthopedic devices, material processing, and manufacturing technologies.

Plastic state of matter is of interest to many branches of science and engineering. The scientific study of the plasticity of metals is regarded to have its beginning in 1864. In that year Tresca published a preliminary account of experiments on punching and extrusion, which led him to state that a metal yielded plastically when the maximum shear stress attained a critical value. Tresca's yield criterion [68] was applied to determine the stresses in a partly plastic cylinder subjected to torsion or bending and in a completely plastic tube expanded by internal pressure.

A review of the development of the theory of elasticity shows that during the period of two hundred years the theory has gradually developed into an important part of mechanics, which is essential for the solid foundation for the design of engineering structures. There are many modern developments in the subject that have resulted in the development of criterion for new engineering constructions and devices that operate under elevated or low-temperature conditions. The influence of elevated temperature on material properties has been considered in the design of steam turbines, automobiles parts, and in oil refining and other chemical equipments. In recent year's development in the design of high-speed aircraft, gas turbines, missiles, rockets, and nuclear reactors have been among the developments in which high temperatures exist and which have found the relevance of such studies. In these applications, the influence of high temperature on the material properties is an important design consideration. The extent of this influence depends upon many factors including the material loading conditions and state of stress.

Moreover elevated temperatures produce creep or "time-flow" in materials. Creep can be defined as the time-dependent deformation produced in solids subjected to stress. For many materials, including most metals and alloys, elevated temperature must be applied for creep to be produced. The first recorded experiments relating to creep appeared in 1830's in connection with suspension bridges, measuring instruments and steam engines.

In 1929, Norton discovered the exponential law $\dot{\varepsilon} = \frac{d\varepsilon}{dt} = K\sigma^n$; where $\dot{\varepsilon}$ is strain rate, σ is stress, K and n are constants. This applies to many metals. Around 1930's, Bailey showed that creep deformation of structural metals (as in time-dependent plasticity) takes place under constant volume and a superimposed hydrostatic pressure does not influence creep deformation. From these facts and the assumption of isotropy, Odqvist [40] deduced constitutive relations for secondary creep under triaxial stresses. These have the same form as von Mises equation for time-independent plasticity. A lot of progress has been made theoretically as well as experimentally since then to define certain aspects of the involved mechanisms of creep. However, no complete theory to explain this complex creep phenomenon is available yet.

8.2 ELASTIC-PLASTIC AND CREEP PHENOMENON

Theory of elasticity deals with the systematic study of stress, strain, and displacement in an elastic body under the influence of external forces. All structural materials posses to a certain extent the property of elasticity, that is, if external forces producing deformation of a structure do not exceed a certain limit, the deformation disappears with the removal of the forces. For various engineering disciplines, the purpose of studying elasticity is to analyses the stresses and displacements of structural and machine elements in this range and to thereby check the sufficiency of their strength, stiffness, and stability. The theory of plasticity is the name given to the mathematical study of stress and strain in plastically deformed solids. This follows the well-established precedent set up by the 'theory of elasticity.' The relation of plastic and elastic properties of metals to crystal structures and cohesive forces belongs to the subject now known as 'metal physics.' The theory of plasticity takes, as its starting point, certain experimental observations of the macroscopic behaviors of a plastic solid in uniform states of combined stress. The task of the theory is twofold: First, to construct explicit relations between stress and strain agreeing with the observations as closely and as universally as needed and second, to develop mathematical techniques for calculating non-uniform distributions of stress and strain in bodies permanently distorted in any way. Unlike elastic solids in which the state of strain depends only on the final state of stress, the deformation that occurs in a plastic solid is determined by the complete history of loading. The plasticity problem is, therefore, essentially incremental in nature, the final distortion of the solid being obtained as the sum total of the incremental distortions following the strain path. The situation in which material deforms continuously with load for a prolonged period of time, (usually at elevated temperature) is called 'creep' and so a constant load test is called a 'creep test.' The conventional stress (load divided by initial cross-section) is called the 'creep-stress.' The gradual strain is called 'creep-strain,' the strain-time curve one obtains is called the 'creep-curve' and the slope of this curve is called the 'creep-rate.' Creep strains may be elastic, plastic or a combination of these. Creep strain might occur too slowly to detect or too rapidly to follow. Even in such cases, however, there is evidence that creep strain is always time-dependent. For many materials including most metals and alloys, elevated temperatures may be applied for creep to be produced. Changes in temperature induce thermal expansion and have ill-defined effects upon the material behavior. However, in various non-metallic materials such as plastics, wood, and concrete, creep occurs at normal temperatures.

Theory of elasticity and plasticity describe the mechanics of deformation of most engineering solids. Both the theories as applied to metals and alloys are based on experimental studies of the relation between stress and strain in a polycrystalline aggregate under simple loading conditions. Thus, they are of phenomenological nature on the macroscopic scale and owe little to the knowledge of the structure of metals. However, in order to understand the limitations so imposed on the theories, an engineer, with his main interest in design and manufacture, must have some knowledge of the structure of metals. The validity of the predictions of any mathematical model of deformation and stress-distribution, no matter how carefully constructed, depends ultimately on one's ability to assess the mechanical properties and characteristics of the material being used and how these data have been applied in the actual calculations. For this reason, testing material properties is a matter of utmost importance for an engineer. No design can possibly be regarded as successful unless it is firmly based on a good understanding of the potential of the material and its consequent ability to respond satisfactorily to the proposed system of loading. It is important to realize that any mechanical testing technique, no matter how appropriate, can provide information only about the average material properties to be generally expected and cannot provide an explanation of why a material behaves in a certain manner. It does not even necessarily follow that data obtained in a test on a specific material will be totally reliable when the material is ostensibly the same, but from a different batch that has been used in the manufacturing process. We can only expect that in the absence of any unknown internal defects and metallic and non-metallic inclusions, the alloy will behave in the way suggested by the test results. The basic method for bringing out the behavior of structural materials is the tensile test.

In tensile test, in which cylindrical specimens of uniform cross-section are extended longitudinally, elastic behavior described by Hooke's law. $\sigma = E\,e$, holds provided the strain e remains always below a limiting value e^*, i.e., $\leq e^*$ where e^*, characteristic of the metal in question, is at the most of the order 5×10^{-3} (for the hardened steels) and may be as small as 10^{-3} (for annealed metals). For strains in the excess of e^* the typical behavior of a metal is shown in Figure 8.1. Immediately beyond e^* the slope of stress-strain curve decreases slightly and this behavior persists up to a value e_y where e_y is slightly larger then e_y. For strain cycles where the strain never exceeds e_y, the mechanical behavior of the material is defined by the stress-strain diagram. For strain cycles where the strain exceeds e_y, permanent deformation is ensured so that even when the specimen is unloading to the stress-free state permanent strain (known as plastic strain) remains. For strain cycles where the strain exceeds e_Y, the detailed stress-strain behavior is defined in Figure 8.1. In the case where e increases continuously, the stress increases monotonically with strain, but usually with a monotonically decreasing slope up to the point of complete failure e_f, where the material fractures and thereby terminates the testing procedure. The quantities e^*, e_Y and e_f are usually known respectively as the limit of proportionality, the yield strain and the fracture strain. Associated with these quantities are the corresponding stresses σ^*, σ_Y is the yield stress and σ_f is the fracture stress. The solid curve of stress vs. strain up to $e = e_f$ is known simply as a stress-strain curve.

Values of e_f vary for the different metals up to a maximum of about 0.4. In the experimental curve (like the one of Figure 8.1) the strain is usually defined as $\left(l - l_0 / l_0 \right)$ where l is the length of deformed specimen originally of length l_0. In experiments involving large strains (say up to $e = 0.4$) the experimental results are plotted using the measures virtually coincide, as for example in the initial elastic region. For the strain cycles where e increases up to a maximum of e_1, which lies between e_Y land e_f, and then subsequently decreases until the stress-free state is attained, the stress- distribution path follows the stress-strain curve up to e_1. During the unloading part of the cycle, the stress-strain locus is shown as the dotted straight line which is a line of the slope of E, parallel to the initial elastic slope, and passing through e_1. On complete unloading to the stress-free state $\sigma = 0$, the specimen is left with a permanent or residual plastic strain e_p as indicated in Figure 8.1. On subsequent reloading of the specimen, the path followed is the dotted lines till the stress-strain curve is reached. After

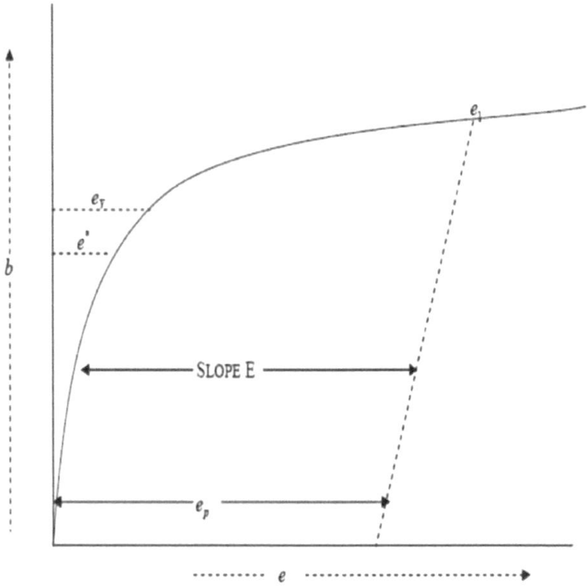

FIGURE 8.1 Behavior of metals in uniaxial tension.

that whereupon the stress-strain locus again joins the latter curve either to fracture or to a point similar to, which is the maximum strain for the new cycle. A completely characteristic feature of metal stress-strain curves is that the (varying) slope in the plastic region is much smaller than E. The stress-strain curve is characteristic of the material and depends not only on the chemical constitution but also on heat-treatment and methods of manufacture.

The strain at fracture is used as a measure of ductility, i.e., if the fracture strain is 'large,' the material is said to be ductile and if it is 'small,' it is said to be brittle. 'Large' and 'small' are usually related to the strain at yield. The stress-strain behavior that is observed namely, the initial elasticity followed by plasticity to fracture, does not depend upon time. As soon as the stress changes, the strain changes. Elastic and plastic behavior are thus said to be instantaneous. However, they depend on the rate of straining. An increase in the rate of straining of the material, irrespective of the method of testing, increases the strain hardening effect and therefore raises the level of the yield stress. That is if the tensile test is performed at different strain rates from very slow to very fast, the typical effect on the stress-strain curve is shown in Figure 8.2 which shows that an increase in strain rate causes an increase in elastic modulus, yield stress and fracture stress and decrease in ductility. The material properties at high strain rates depend not only on the strain actually imposed but also, to a larger degree than in the static conditions, on the strain hardening effects and therefore on the thermal control which influences the movement of dislocations and the rate of their flow. Usually, an increase in the test temperature decreases both the yield and elastic modulus. The temperature also influences the fracture ductility.

It is also observed that at high-temperature plastic strain under the effect of a relatively small stress grows with time. This phenomenon, which is called creep, is expressed in certain cases by strains that increase with time, while the load remains constant, and in some other cases by the continuous decrease of stress while the strain remains constant (relaxation). Creep determines the resistance and the duration of mechanical elements submitted to high temperatures. In general, creep-time curves are of the shape as shown in Figure 8.1 where four stages in the creep time variation can be distinguished. At $t = 0$ the curve OA shows an instantaneous response ε_0 which, depending on the magnitude of the stress, could be elastic or elasto-plastic. The portion AB is characterized

ε by a relatively high creep strain rate which decreases with time as a result of strain hardening. This is called primary creep, transient creep or logarithmic creep.

This portion of the curve is attained at both low and high temperatures. In fact, it can occur in the absence of thermal activation also; for example, at 4° K. The portion BC, which is linear, corresponds to constant creep rate or maximum creep rate where the effect of strain hardening is balanced by an annealing influence and is called the secondary creep.

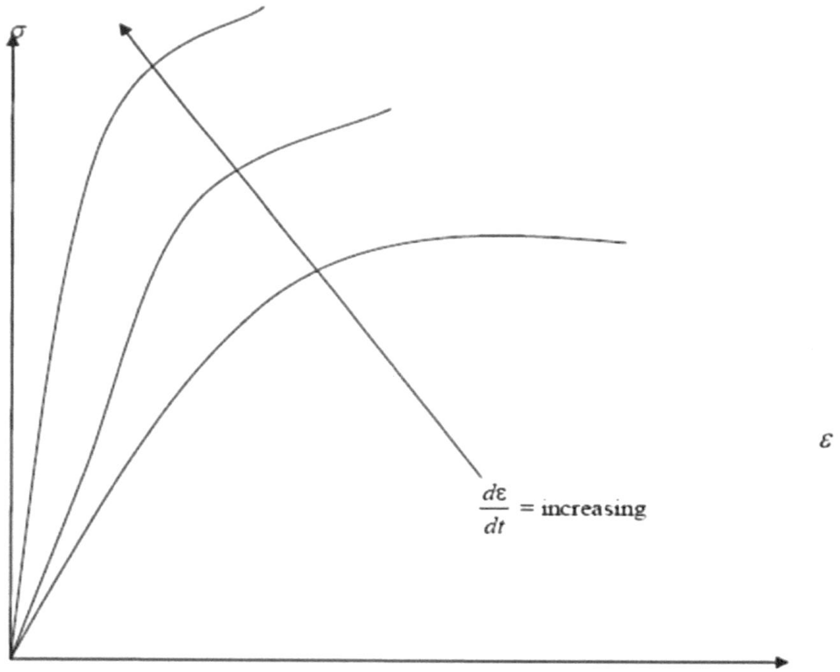

FIGURE 8.2 Effect of strain rate on tensile.

At some point, another transition occurs and creep rate again accelerates and an upward curving portion CD is attained. This final stage, which is called tertiary creep, represents a region where the creep rate continues to increase and where the reduction in cross-sectional area is accompanied by an increase in stress resulting in fracture. For lower stresses and temperatures, the final stage of creep is not observable during the usual times covered by creep tests. Materials differ in the arrangements of these regimes. Some have hardly any secondary creep while others have hardly any tertiary creep and so on. The curve also shows that for every stress rupture eventually occurs.

We usually associate creep with high temperatures but whether the temperature required is high or not, it really depends upon the material. Potentially, creep can take place at any temperature above absolute zero. In various non-metallic materials such as concrete, wood, and natural high polymers, creep occurs at normal temperatures [6]. For typical structural materials like magnesium (Mg), lead (Pb) and its alloys, creep occur at or below room temperature (RT) [39] whereas in other materials such as metals, creep occurs at high temperatures.

Many structural models [3, 4, 9, 31, 33, 34, 36, 59, 61] have been limited to isotropic homogeneous materials. In the strength of the material of the body being analyzed, there are many important assumptions such as material is continuous, homogeneous, and isotropic. A continuous body is one that does not contain voids or empty spaces of any kind. A body is homogeneous if it has identical properties at all points. A body is considered to be isotropic with respect to some property when that property does not vary with direction or orientation. Real bodies can possess an initial non-homogeneity due to the inclusion of a foreign material or as a result of being a composite material. Non-homogeneity can also be generated by some external fields such is the thermal field. The elastic modulus of the material may vary with the temperature. Effect of including non-homogeneity on stress distribution caused by the external field is much more pronounced and is of longer duration than the effect of thermal stresses themselves. The interplay between the effect of homogeneity and inhomogeneity is of particular importance in the plastic zone of deformation because the additional straining of the material, caused by macro shears affects the value of the yield stress. The amount of homogeneous deformation does not depend on the mechanical properties of the material formed but only on the geometry of the system in which the flow is taking place however inhomogeneous deformation is affected by both the geometry of the system, i.e., by the type of flow pattern induced and by characteristics of stress-strain curve of the material in question (Figure 8.3).

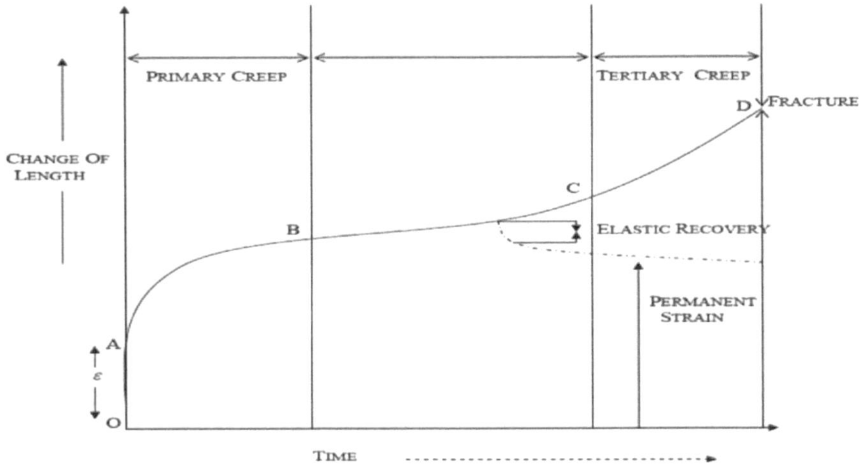

FIGURE 8.3 Creep curve for three stages.

Relatively fewer investigators [14, 20, 28, 29] of stress and strain in non-homogeneous bodies have appeared in the literature. This is due, in part, to the mathematical complexity of such problems and to the fact that engineering structures have generally been fabricated from materials which are essentially isotropic. However, because of recent aerospace and commercial applications such as helicopter, rotor blade, compressors, flywheel, automobile structures, etc., anisotropic non-homogeneous materials having variable thickness are being effectively utilized. Thus, there is an obvious need for further research in this area. A number of review papers, books, and proceedings of symposia have been published with macroscopic elastic, plastic, and creep behavior of homogeneous and non-homogeneous materials. The technical notes include those by Jiang in the year 1992 and the proceedings of the symposia edited by Dorn [6], Odquist [38], Penny, and Mariotte [41] and Robotnov [44]. The books published on the subject include those by Boyle and Spence [2], Chakarbarty [18], Fung [8], Hill [31], Johnson, and Mellor [33], Lubhan, and Felgar [38], Nadai [37], Odquist [38, 39], Parkus [40], Sokolnikoff [59], Timoshenko, Goodlier [60], and Zvolinsky [63].

8.3 FINITE DEFORMATION

There are many technically important problems in elasticity including those of buckling and stability that call for a consideration of finite deformation in which the displacements together with their derivatives are no longer small. Theory of finite strains provides an illustration of the complications that appear in the development of a theory when fundamental equation becomes non-linear. In recent years a good deal of work has been done on the theory of finite deformations [7, 28, 29, 48–57]. It has been applied to various problems which cannot be dealt with classical theory of small deformation (infinitesimal). It has qualitatively predicted a yield point and has also combined into one the rival hypotheses of elastic failure (i.e., principal maximum stress hypothesis and maximum shear stress or maximum principal stress difference hypothesis). It has proved the existence of Bauschinger effect by providing a result that yield stress in compression can be several times greater than in tension. Its application gives axial stresses in cylinders subjected to large torsional shafts, which were earlier neglected in the classical theory. To define definite finite deformation in a continuous medium, there are two methods, namely, the Lagrangian and the Eulerian. Components of strain can be described by the co-ordinates of a typical particle either in the strained state as independent variables or in the unstrained state as the independent variables. The former is known as the Eulerian description and the later the Lagrangian description. Many investigators [1, 5, 42] have adopted the Lagrangian view point for reasons of mathematical convenience whereas the importance of using the Eulerian method has been stressed by Seth [55], Ritz [47], and Zvolinsky [73]. In this connection the following note from the paper by Seth [56] is worth quoting: "Like the body stress equations, these (strain components) should be referred to the actual position of a point P of the material in the strained condition, and not to the position of a point considered before strain. The importance of this point, overlooked by various authors, cannot be exaggerated.

Now we consider an aggregate of particles in a continuous medium. Let the coordinates of a particle lying on a curve c_0 (before deformation) be denoted by (A_1, A_2, A_3) and let the co-ordinates of the same particle after deformation (which is lying on the curve c) be (x_1, x_2, x_3). Then the elements ds_0 and ds of the arc of the curves c_0 and c, respectively, are given by

$$ds_0^2 = dA_1^2 + dA_2^2 + dA_3^2 = dA_i dA_i \ (i = 1, 2, 3) \tag{1}$$

and

$$ds^2 = dx_1^2 + dx_2^2 + dx_3^2 = dx_i dx_i \ (i = 1, 2, 3) \tag{2}$$

In Eulerian description of strain, we have

$$A_i = A_i\left(x_1, x_2, x_3\right) \tag{3}$$

Substituting Eq. (3) in Eq. (1), we have

$$ds_0^2 = dA_i dA_i = A_{i,j} A_{i,k} dx_j dx_k \ (i, j, k = 1, 2, 3) \tag{4}$$

and

$$ds^2 = dx_i dx_i = \delta_{jk} dx_j dx_k \tag{5}$$

We have the necessary and sufficient condition for the transformation $A_i = A_i\,(x_1, x_2, x_3)$ to be one of rigid body motion in that ds^2 and ds_0^2 should be equal for all the curves c_0. Hence, we take the difference $ds^2 - ds_0^2$ as a measure of strain and write

$$ds^2 - ds_0^2 = 2\overset{A}{e}_{jk}\, dx_j dx_k \tag{6}$$

The strain tensor $\overset{A}{e}_{jk}$ was introduced by Cauchy for infinitesimal strains, by Almansi and Hamel for finite strain and is known as Almansi strain tensor. Subtracting (4) from (5), we get

$$ds^2 - ds_0^2 = \left(\delta_{jk} - A_{i,j} A_{i,k}\right) dx_j dx_k \tag{7}$$

Now comparing Eq. (6) and Eq. (7), we get

$$2\overset{A}{e}_{jk} - ds_0^2 = \delta_{jk} - A_{i,j} A_{i,k} \tag{8}$$

Writing strains $\overset{A}{e}_{jk}$ in terms of displacement components $u_i = x_i - A_i$; we get

$$2\overset{A}{e}_{jk} = \left[u_{j,k} + u_{k,j} - u_{i,j} u_{i,k} \right] \tag{9}$$

where the functions $\overset{A}{e}_{jk}$ are called the Eulerian strain components.

If we use Lagrangian coordinates, then A_i are taken as independent variables and equations of transformation are of the form

$$x_i = x_i \,(A_1, A_2, A_3)$$

We can also write

$$dx_i = x_{i,j} da_j \;\; (i, j = 1, 2, 3) \tag{10}$$

Thus the elements ds_0 and ds of the arc of the curve c_0 and c are given by

$$ds_0^2 = dA_i dA_i = \delta_{jk} dA_j dA_k \tag{11}$$

and

$$ds^2 = dx_i dx_i = x_{i,j} x_{i,k} dA_j dA_k \tag{12}$$

As a result, the Lagrangian components of strain ε_{jk} are defined as

$$ds^2 - ds_0^2 = 2\varepsilon_{jk} dA_j dA_k \tag{13}$$

Also from Eqs. (11) and (12), we have

$$ds^2 - ds_0^2 = \left(x_{i,j} x_{i,k} - \delta_{jk} \right) dA_j dA_k \tag{14}$$

Now comparing Eqs. (13) and (14), we get

$$2\varepsilon_{jk} = x_{i,j} x_{i,k} - \delta_{jk} \tag{15}$$

Expressing ε_{jk} in terms of displacement components $u_i = x_i - A_i$, it becomes

$$2\varepsilon_{jk} = u_{j,k} + u_{k,j} + u_{i,j}u_{i,k} \quad (i, j, k = 1, 2, 3) \tag{16}$$

To show the fact that the differentiation in Eq. (9) is carried out with respect to the variable x_i (strained coordinates) while in Eq. (16) it is carried out with respect to variable A_i (unstrained co-ordinates), the typical expression for $\overset{A}{e}_{jk}$ and ε_{jk} in unabridged form can be written as:

$$\overset{A}{e}_{xx} = \frac{\partial u}{\partial x} - \frac{1}{2}\left[\left(\frac{\partial u}{\partial x}\right)^2 + \left(\frac{\partial v}{\partial x}\right)^2 + \left(\frac{\partial w}{\partial x}\right)^2\right]$$

$$\varepsilon_{aa} = \frac{\partial u}{\partial a} - \frac{1}{2}\left[\left(\frac{\partial u}{\partial a}\right)^2 + \left(\frac{\partial v}{\partial a}\right)^2 + \left(\frac{\partial w}{\partial a}\right)^2\right]$$

and

$$2\overset{A}{e}_{xy} = \frac{\partial u}{\partial y} + \frac{\partial v}{\partial x} - \left[\frac{\partial u}{\partial x}\frac{\partial u}{\partial y} + \frac{\partial v}{\partial x}\frac{\partial v}{\partial y} + \frac{\partial w}{\partial x}\frac{\partial w}{\partial y}\right]$$

$$2\varepsilon_{ab} = \frac{\partial u}{\partial b} + \frac{\partial v}{\partial a} - \left[\frac{\partial u}{\partial a}\frac{\partial u}{\partial b} + \frac{\partial v}{\partial a}\frac{\partial v}{\partial b} + \frac{\partial w}{\partial a}\frac{\partial w}{\partial b}\right]$$

where $\overset{A}{e}_{xx}$, ε_{aa} represent the extension of vectors originally parallel to the co-ordinate axis, while $\overset{A}{e}_{xy}$, ε_{ab} represent shear or change of angle between vectors originally at right angles.

For small deformations, that is, if the displacements and their derivatives are small, the two representations, namely Eulerian and Lagrangian, become identical and hence it is immaterial which system is employed. In particular, if we neglect the non-linear terms in the partial derivatives in Eqs. (9) and (16), then the both sets of formulae reduce to:

$$2\overset{A}{e}_{jk} = \varepsilon_{jk} = \frac{1}{2}\left[u_{j,k} + u_{k,j}\right] \quad (j, k = 1, 2, 3)$$

which is valid in the case of infinitesimal transformations. The strain components in cylinder polar coordinates are given by (9)

$$\overset{A}{e}_{rr}=\frac{\partial u}{\partial r}-\frac{1}{2}\left[\left(\frac{\partial u}{\partial r}\right)^2+\left(\frac{\partial v}{\partial r}\right)^2+\left(\frac{\partial w}{\partial r}\right)^2-v^2\right]$$

$$\overset{A}{e}_{\theta\theta}=\frac{1}{r}\frac{\partial v}{\partial \theta}+\frac{u}{r}-\frac{1}{2r^2}\begin{bmatrix}\left(\frac{\partial u}{\partial \theta}\right)^2+r^2\left(\frac{\partial v}{\partial \theta}\right)^2+\left(\frac{\partial w}{\partial \theta}\right)^2-vr^2\frac{\partial u}{\partial \theta}\\ +u\frac{\partial v}{\partial \theta}-v\frac{\partial u}{\partial \theta}+ur^2\frac{\partial v}{\partial \theta}+u^2+r^2v^2\end{bmatrix}$$

$$\overset{A}{e}_{rr}=\frac{\partial w}{\partial z}-\frac{1}{2}\left[\left(\frac{\partial w}{\partial z}\right)^2+r\left(\frac{\partial v}{\partial z}\right)^2+\left(\frac{\partial w}{\partial z}\right)^2\right]$$

$$\overset{A}{e}_{r\theta}=\frac{1}{2}\left[\frac{1}{r}\frac{\partial u}{\partial \theta}+\frac{\partial v}{\partial r}-\frac{u}{r}\right]-\frac{1}{2r}\begin{bmatrix}\frac{\partial u}{\partial r}\frac{\partial u}{\partial \theta}+r\frac{\partial (rv)}{\partial r}\frac{\partial v}{\partial \theta}+\frac{\partial w}{\partial r}\frac{\partial w}{\partial \theta}\\ -vr^2\frac{\partial u}{\partial r}+\frac{u}{r}\frac{\partial (rv)}{\partial x}-rv\frac{\partial v}{\partial \theta}\end{bmatrix}$$

$$\begin{aligned}\overset{A}{e}_{\theta z}&=\frac{1}{2}\left[\frac{\partial v}{\partial z}+\frac{1}{r}\frac{\partial w}{\partial \theta}\right]-\frac{1}{2r}\left[\frac{\partial u}{\partial r}\frac{\partial u}{\partial \theta}+r^2\frac{\partial v}{\partial z}\frac{\partial v}{\partial \theta}+\frac{\partial w}{\partial z}\frac{\partial w}{\partial \theta}-v\frac{\partial u}{\partial z}+r^2u\frac{\partial v}{\partial z}\right]\\ \overset{A}{e}_{zr}&=\frac{1}{2}\left[\frac{\partial w}{\partial r}+\frac{\partial u}{\partial z}\right]-\frac{1}{2}\left[\frac{\partial u}{\partial z}\frac{\partial u}{\partial r}+r\frac{\partial v}{\partial z}\frac{\partial (rv)}{\partial \theta}+\frac{\partial w}{\partial z}\frac{\partial w}{\partial \theta}+rv\frac{\partial v}{\partial z}\right]\end{aligned} \tag{17}$$

Here u, v, w and $\overset{A}{e}_{rr}, \overset{A}{e}_{\theta\theta}, \overset{A}{e}_{zz}, \overset{A}{e}_{r\theta}, \overset{A}{e}_{\theta z}$ and $\overset{A}{e}_{zr}$ are the physical components of the displacement u_i and strain tensor $\overset{A}{e}_{ij}$, respectively. In the case of disc/cylinder, the distribution of stress is symmetrical about the axis of revolution, and as a result, the stresses and strains are independent of polar angle θ. The displacement components in cylindrical polar coordinates are given by [52]

$$u = r(1-\beta),\ v = 0 \text{ and } w = dz \tag{18}$$

where β is a function of $r=\sqrt{x^2+y^2}$ only and d is a constant.

Substituting (18) in (17), the finite components of strain are:

$$\begin{aligned}
\overset{A}{e}_{rr} &\equiv \frac{\partial u}{\partial r} - \frac{1}{2}\left(\frac{\partial u}{\partial r}\right)^2 = \frac{1}{2}\left[1-\left(r\beta'+\beta\right)^2\right] \\
\overset{A}{e}_{\theta\theta} &\equiv \frac{u}{r} - \frac{1}{2}\frac{u^2}{r^2} = \frac{1}{2}\left[1-\beta^2\right] \\
\overset{A}{e}_{zz} &\equiv \frac{\partial w}{\partial z} - \frac{1}{2}\left(\frac{\partial w}{\partial z}\right)^2 = \frac{1}{2}\left[1-\left(1-d\right)^2\right] \\
\overset{A}{e}_{r\theta} &= \overset{A}{e}_{\theta z} = \overset{A}{e}_{zr} = 0
\end{aligned} \tag{19}$$

where

$$\beta' = \frac{d\beta}{dr}$$

8.4 TRANSITION

Transition is a natural phenomenon and there is hardly any branch of science or technology in which we do not come across the transition from one state into another. Elasticity-plasticity, visco-elastic, deformations, creep, fatigue relaxation, and adiabatic theory in quantum mechanics are well-known examples. At present, these are generally treated as discontinuous. For elastic-plastic deformations, the ideal model has to assume some type of yield condition. For creep, not only the yield condition but also a number of creep strain laws have to be assumed. All these assumptions leave one a little dissatisfied. More scientific advancements should attempt to reduce these assumptions. This has motivated us to pay some attention to the treatment of transition problems in the domain of solid mechanics.

The deformation theory of plasticity does not satisfy the condition of continuity of the relationship between stress and deformation on the transition from loading to unloading. In analyzing this and other theories of plasticity, we may encounter similar circumstances in which certain consequences of the theory appear to be contradictory or physically unacceptable. In the classical theory of plasticity, the material region is assumed to be divided into elastic and plastic regions, which are separated by a yield surface depending on the symmetry and other physical considerations. In other words, perfect elasticity and ideal plasticity are two extreme properties of the material and the use of ad-hoc rule like yield condition amounts to dividing the two extreme properties by a sharp line which is not physically possible. In the elastic region, the classical theory of linear elasticity holds. In the plastic region, Prandtl-Reuss or Von-Mises equations are used with a yield condition and the boundary condition that the normal stress components should be continuous on the yield surface. This linear theory has given rise to a vast amount of important results which have been corroborated by experiments more accurately than one may expect. However, its main drawbacks are the following assumptions:

- That even through the material at a point has yielded; the material at a neighboring point still remains elastic.
- That a yield surface of the assumed type separating the elastic and plastic regions exists.
- That for any given material there is a function of the three principal stresses which always has a value when yielding begins regardless of the stress state.
- That the same functional relationship applies to all materials although the numerical value of the function is different for different materials.

The elastic-plastic transition has been obtained in current literature with the help of semi-empirical yield condition like that of Tresca or Von-Mises. The stresses are obtained from the elastic solution and then substituted in the yield condition to get the transition surface. The possibility of treating it as a transition or turning point phenomenon in finite deformation has not been explored. When plastic state tends to set in, the stress-strain relation undergoes a change. This must be reflected in our equations. The linear classical theory cannot do it. Tresca [68] pointed out that a transition state (which he called mid-zone) exists when a material passes from elastic-zone to plastic-zone and this was later on supported by Todhunter and Pearson. In the transition state whole of the material participates, and not simply a selected region or a line as assumed by classical theories. A recent numerical study [10, 11] on the flow and deformation theories in plasticity was undertaken to see as to what extent a continuous approximation involving the idea of transition to an elastic-plastic material in terms of the stress-strain law, would lead to a satisfactory convergent solution. The results obtained showed excellent approximations and convergence to elastic-perfectly plastic solution.

The demand of high-speed technology in transportation, communication, and energy conversion has forced researchers to take serious note of non-linearity. However, some still find it difficult to get rid of the century-old habit of analytical and experimental research in continuum mechanics, which leaned very heavily on linearization. If a medium 'A' changes into '*B*' through a transition, state *T*, '*A*' and '*B*' may be almost linear but T is non-linear. Since this non-linearity is difficult to investigate, workers have taken to the artifice of replacing it by singular, non-differentiable, or discontinuous surfaces. This piece-wise treatment necessitates the use of ad-hoc and semi-empirical laws, which may or may not exist. Linearizing non-linear problems by perturbation, boundary layer, and other techniques do not provide a satisfactory explanation for some important characteristics of non-linearity. As a result, a number of important physical effects do not get an adequate scientific expression. There is hardly any information, which we do not know about any linear field. Their existence, uniqueness, and stability are well established. Nature does not always conform to our abstract concepts of linearity, smoothness, symmetry, identity, homomorphism, and isotropy. It is true that physical phenomenon tends to behave linearly in the course of time, which may be millions of years but the demands of modern technology want to compress years into a fraction of seconds. Thus transition, which frequently occurs in nature, has to be tackled. In fact, all linear-disciplinary fields, which are so important in modern research, give rise to important transition problems. But both the macro and micro analysts have devoted very little attention to them due to non- linearity involved in the analytic treatment. There are four different ways of treating transition fields. Firstly, they are asymptotic in character and hence they should be associated with some singularities or criticalities of the differential system describing them. If the singularities are not obvious in one plane, it should be possible by continuous mapping to recognize them in some other plane. Secondly, the transition field may be interpreted as asymptotic sub-spaces obtained from the intersection of two spaces representing different media. Thirdly, from the group-theoretic point of view, all continuous deformations form a symmetric group.

At transition, the nature of this group changes. For example, an elastic body, which belongs to the orthogonal group, becomes uni-molecular on becoming plastic. Lastly, from the macro point of view, one can imagine that at transition the macro-element breaks down, with the result that the corresponding transformation matrix becomes singular.

In general, the material from the elastic state can go over into:

- plastic state;
- to creep state; or
- first to plastic then to creep or vice-versa under external loading system.

When the material under the experiment goes from a primary state of creep to secondary or to tertiary states and, from secondary to tertiary state, the transition takes place. A plastic or a creep state is a transition state from an initially elastic state. All these cases of transition can be expected to occur [48, 49, 55–57] when some functions of elasticity of the medium take on critical values. These functions are called transition functions. These may be either principal stresses, or principal stress differences, or stress invariants, or maybe any suitable combination of these. These critical (asymptotic) values have to be determined at the transition points of the differential equations describing the medium. If a number of transition states occur at the same point, the transition function will have different limiting values, and the point will be a multiple one, each branch of which will then correspond to different states, whereas the classical treatment uses different constitutive equations for each state involved.

8.5 GENERALIZED STRAIN MEASURE

The response of real materials to the external loading, in general, is nonlinear in character. The division of deformation into different types arose from a desire for linearization of engineering problems. In large deformations, which involve plastic flow, creep, and fatigue, the current treatment requires a number of ad-hoc and semi-empirical assumptions. The adoption of these semi-empirical laws has complicated the problem without evolving any overall simple concept governing them. One source of troubles is the use of classical measures of deformation produced in a medium even when we know from experiments that non-linearity is a characteristic of such deformed media. Although now abstract measures have been highly developed, these have as yet not been employed in the problems of non-linear mechanics suitably. In classical mechanics, ordinary measures have been found sufficient and there arises no need for their extension. The equation of equilibrium and the concept of stresses are well defined; only the measures of deformation are flexible. If we place any restriction on these measures, the constitutive equations will assume complicated forms, even in the cases where we feel, it should be done. Plastic and creep effects are well-known examples. In such cases, the strain is small and no modification of constitutive equations can better the results. Such problems are treated as a transition phenomenon. A continuum approach necessarily means the introduction of non-linear measures. Deformation fields associated with an irreversible phenomenon such as elastic-plastic deformation, creep, relaxation, fatigue, fracture, etc. have been known to be non-linear in character as revealed by extensive experimental studies. The use of classical measures of deformation is totally inadequate to deal with their transitions and hence the corresponding constitutive equations of the medium have to be made complicated. From the above analysis, it appears that one should construct a generalized measure of deformation to resolve the difficulty of using classical measures to explain the natural phenomenon in continuum mechanics.

The nature of the strain rate described in four stages of creep- elastic, transient, secondary, and rapture shown in Figure 8.3 is different in each case. As the deformation is non-linear and hence there arises the need for generalized strain rate measure which can be used in all the stages. Robotnov [44, 45] has pointed out difficulties; particularly the ambiguities in the interpretation of the experimental data that are involved in the choice of suitable constitutive equations for various states of creep described in Figure 8.3. Two parameters characterize each of these states, one for the measure and other for the irreversibility. Since the creep strain rate depends upon the stresses and temperature at a structural state, it is not the total strain but the total rate of creep strain, which is significant. Therefore, it is expected that a generalized measure concept in which the two parameters are experimentally determined may give a better insight into creep behavior. Seth [49] defined the generalized principal strain measure as:

$$\overset{A}{e}_{ii} = \int_0^{\overset{A}{e}_{ii}} \left(1-2\overset{A}{e}_{ii}\right)^{\frac{n}{2}-1} d\overset{A}{e}_{ii} = \frac{1}{n}\left[1-\left(1-2\overset{A}{e}_{ii}\right)^{n/2}\right] \tag{20}$$

Here $\overset{A}{e}_{ii}$ are principal Almansi finite strain components. In the Cartesian framework, we can readily write down the generalized measure in terms of any other measure. For the uni-axial case, it is given by Seth [49] as:

$$e = \frac{1}{n}\left[1-\left(\frac{l}{l_0}\right)^n\right] \tag{21}$$

where l_0 and l are the initial and strained lengths of the rod respectively. For $n = -2, -1, 0, 1, 2$, it gives the Cauchy, Green, Hencky, Swainger, and Almansi measures respectively. Seth [49, 52, 56] has shown that the well-known creep strain laws used in current literature such as Norton's law, Kachonov law, Odqvist law, Andrade's law, etc. can be derived from the generalized measure.

The generalized strain measure not only gives the well-known strain measures as special cases, but it can also be used to find the creep stresses when it is combined with the transition point analysis of the governing differential equations. Seth has shown that the transition point analysis does not require the assumption of incompressibility, creep strain law and yield condition as used by many authors [9, 39, 62]. He, as well as others, have successfully applied it to a number of problems [11–27, 32, 46–57] showing that the asymptotic solution of the governing differential equations at the transition point gives the results which are obtained by assuming yield criteria when they exist. The most important contribution that can be made by generalized measure is that it will make use of semi-empirical laws and jump conditions unnecessary. If such a law exists, then these come out of the analytic treatment as a particular case. Thus, an important function of non-linear measures is to explain transition without assuming conditions to match the two solutions at transition.

8.6 CONSTITUTIVE EQUATIONS

There are many equations characterizing the individual material and its reaction to applied loads. These are called constitutive equations since they describe the macroscopic behavior resulting from the internal constitution of the material. But materials, especially in the solid-state, behave in complex ways when the entire range of possible temperatures and deformations is considered. As a result, it is not feasible in classical theory to write down one equation or a set of equations to describe accurately a real material over its entire range of behavior. Instead, separate equations are formulated to describe the various kinds of ideal material response, each of which is a mathematical formulation designed to approximate physical observations of a real material's response over a suitably restricted range. The classical equations were introduced to meet specific needs separately and were made as simple as possible simplifying many physical situations. Some of the ideas involved in formulating simple equations for such ideal materials are illustrated in subsections.

8.6.1 ELASTIC STATE

A material is said to be elastic when a body formed of the material recovers its original form completely upon removal of forces causing the deformation, and there is one to one relationship between the state of stress and state of strain for a given temperature. It is assumed to obey Hooke's law, which for uniaxial stress situations states, "Extension produced by tensile forces is proportional to the force." Under triaxial loading, classical elasticity theory assumes a generalized Hooke's law expressing each stress component as a linear combination of all the strains, that is

$$T_{ij} = C_{ijkl} e_{kl} \; (i, j, k, l = 1,2,3) \tag{22}$$

where T_{ij} and e_{ij} are the stress and strain tensors respectively. These nine equations contain a total of 81 coefficients C_{ijkl} but not all the coefficients are independent. The symmetry of T_{ij} and e_{ij} reduces the number of independent coefficients to 36. For elastically isotropic material, which has the same elastic properties in all directions, coefficients reduce to two independent elastic constants. The generalized Hooke's law for the elastic isotropic material at constant temperature can be written in the following form

$$T_{j}^{'} = 2Ge_{ij}^{'} \tag{23}$$

$$T_{ii} = 3Ke_{ii} \tag{24}$$

where K and G are bulk modulus and shear modulus of the material $T_j^{'}$ and $e_{ij}^{'}$ are the stress deviation and strain deviation respectively, i.e.,

$$T_j^{'} = T_{ij} - \frac{1}{3}T_{ii}\delta_{ij} \tag{25}$$

and

$$e_{ij}^{'} = e_{ij} - \frac{1}{3}e_{ii}\delta_{ij} \tag{26}$$

If we substitutive (25) and (26) into (23) and make use of (24), the result may be written in the form

$$T_{ij} = \lambda\delta_{ij}I_1 + 2\mu e_{ij} \tag{27}$$

where $\lambda = K - \frac{2}{3}G$ and $\mu = G$ are Lame's constants, δ_{ij} is Kronecker's delta and $I_1 = e_{ii}$ is the first strain invariant.

The coefficient of the constitutive equations discussed above specifying the relationship between stress and strain for the material in general depend on the temperature, but we usually assume that the temperature variation is sufficiently small so that the coefficients may be treated as constants during the deformation. Even though we neglect the variation of the elastic constants with temperature, we may have to take into account the thermal expansion of the material, which often produces dimensional changes as large as those produced by the applied forces, or, if the dimensional change is prevented by support constraints or surrounded material, thermal stresses are induced in addition to the stresses related to the strains according to the elastic constitutive equations. The thermoelastic constitutive equations for isotropic material are given by [8, 40],

$$T_{ij} = \lambda\delta_{ij}I_1 + 2\mu e_{ij} - \xi\theta\delta_{ij} \tag{28}$$

where $\xi = \alpha(3\lambda + 2\mu)$; θ being the coefficient thermal expansion and θ the rise of temperature. For the special case of steady heat flow, we have

$$\nabla^2\theta = \theta_{,ii} = 0. \tag{29}$$

8.6.2 PLASTIC STATE

Metals obey Hooke's law only in a certain range of small strain. When a metal is strained beyond an elastic limit, Hooke's law is no longer valid. The behavior of metals beyond their elastic limit is rather complicated as discussed in Section 8.1.2. For analysis of continuum stress and strain distributions, a constitutive theory of plasticity must satisfy the yield condition under combined stresses, since the uniaxial condition $|T| = Y$ is inadequate when there is more than one stress component. We shall now exhibit the minimum ingredients that constitute a theory of plasticity.

For an ideal plastic solid obeying Von-Mises yield criterion and flow rule, the following simplifications are taken:

- The plastic strain assumes incompressibility and the plastic strain deviation tensor is the same as the plastic strain tensor.
- The material is elastic and obeys Hooke's law as long as the second invariant $J_2 = \frac{1}{2}T'_{ij}T'_{ij}$ of the stress deviation tensor is less than a constant K^2. In other words, no change in plastic strain can occur as long as $J_2 < K^2$, i.e., e_{ij}^p when $J_2 < K^2$.
- Yielding can occur (elastic limit is reached) only when $J_2 = K^2$. When the yielding condition $J_2 - K^2 = 0$ prevails, the rate of change of plastic strain is proportional to the stress deviation.

$$e_{ij}^p = \frac{1}{\mu}T'_{ij} \quad ; \mu > 0$$

where μ is a positive factor of proportionality, which has the dimensions of the coefficients of viscosity of a fluid.

- Any stress-state corresponding to $J_2 < K^2$ cannot be realized in the material.

The above set of laws contains two essential parts: the criterion for yielding and stress-strain relations in the elastic-plastic regimes. In these specifications, the yielding condition is based on the second invariant of stress deviation tensor. Such a yielding criterion was first proposed by Von Mises. The constant K could be identified with the yield stress in simple shear. For a work hardening material, K will be allowed to change with strain history. Tresca's yield condition is sometimes used instead of von-Mises yield condition. Tresca through his work on metal forming in an armory concluded that the decisive factor for yielding is the maximum shear stresses in the material. Tresca's criterion stipulates that maximum shear stress must have the constant value K during plastic flow.

To express Tresca's idea analytically, it is convenient to use the principal stresses $T_1 \geq T_2 \geq T_3$. If it is known that $T_1 \geq T_2 \geq T_3$ then Tresca's yielding condition is:

$$f \equiv T_1 - T_2 - 2K = 0$$

However f in this form is not analytic, it violates the rule that the manner in which the principal axes are labeled 1, 2, and 3 should not affect the form of yield function. To obey this rule, we observe that Tresca's condition states that during plastic flow one of the differences $|T_1 - T_2|$, $|T_2 - T_3|$, $|T_3 - T_1|$ has the value 2K. Hence, we may write:

$$f \equiv \left[(T_1 - T_2)^2 - 4K^2\right]\left[\left[(T_2 - T_3)^2 - 4K^2\right]\right]\left[(T_3 - T_1)^2 - 4K^2\right] = 0 \qquad (30)$$

This equation is symmetrical with respect to the principal stresses.

In the above treatment, the ideal theories of elasticity and plasticity are dealt with separately and then linked together through a semi-empirical law, called the yield condition. Such a law may or may not exist. What actually happens is that when a medium starts to yield, a constraint is placed on the invariant of the strain tensor of the field such that it satisfies a functional relation of the form:

$$f(I_1, I_2, I_3) = 0 \qquad (31)$$

where I_1, I_2 and I_3 are the first, second, and third invariants of the strain tensor. The form of f is determined from the condition that the modulus of transformation takes on a singular value like zero or infinity and not by any adhoc considerations. In the current treatment, it is argued that the incompressibility of the material makes I_1 vanish and since I_3 can be taken to be very small; equation (1.31) can be reduced to the form:

$$I_2 = a \text{ constant} \tag{32}$$

which is known as the Huber-von Mises yield condition (29). If two of principal stresses are equal or one is the arithmetic mean of the other two, condition (32) reduces to the Tresca yield condition (30). But it is clear that equation (32) cannot account for the Bauchinger effect, for which I_3 must appear in the yield condition Seth has expressed the yield stress Y in simple tension when the material is in the fully-plastic state:

$$Y = \frac{E}{n} \tag{33}$$

where E is the response coefficient in the transition range. It is also concluded that the yield stress in compression is twice that in tension and the general form of yield condition also contains the Bauchinger effect.

The stress-strain relations for an elastic-perfectly plastic solid were first proposed for the case of plane stain deformation. It was assumed that the plastic stain increment, denoted by a superscript p in the following equations, is at any instant proportional to the instantaneous stress deviation S_{ij} and shear stresses, thus,

$$\frac{de_{11}^p}{S_{11}} = \frac{de_{22}^p}{S_{22}} = \frac{de_{33}^p}{S_{33}} = \frac{de_{12}^p}{T_{12}} = \frac{de_{23}^p}{T_{23}} = \frac{de_{31}^p}{T_{31}} = d\lambda$$
$$\text{Or} \quad de_{ij}^p = S_{ik}\,d\lambda \quad (i, j = 1, 2, 3) \tag{34}$$

where $d\lambda$ is an instantaneous non-negative constant of proportionality which may vary throughout a straining program. Robotnov [49] has pointed out difficulties, particularly the ambiguities in the interpretation of the experimental data and these involved in the choice of suitable constitutive equations for various states of creep described in the Figure 8.3. Two parameters characterize each of these states, one for the measure and other for the irreversibility. Since the creep strain rate depends upon the stresses and temperature at a structural state, it is not the total strain but the total rate of creep strain, which is significant. Therefore, it is expected that a generalized measure concept in which the two parameters are experimentally determined may give a better insight into creep behavior. The equations state that a small increment of plastic stain depends upon the current deviatoric stress, and not on increment, which is required to bring it about. Also the principal axes of stress and plastic stain increments coincide. The equation is only a statement about the ratio of the plastic strain increments in the x, y, z direction; it gives no direct information about their absolute magnitudes. The total stain increment is the sum of the elastic strain increment (denoted by a superscript e) and the plastic strain increment. Thus,

$$de_{ij} = de_{ij}^{p} + de_{ij}^{e} = S_{ij} d\lambda + \left\{ \frac{dS_{ij}}{2G} + \frac{(1-2v)}{E} \delta_{ij} dT_{kk} \right\} \tag{35}$$

where v is Poisson's ratio.

Since the plastic straining causes no change of plastic volume, we may write the condition of incompressibility in terms of the principal or normal strains, as:

$$\begin{aligned} de_{11}^{p} + de_{22}^{p} + de_{33}^{p} &= 0 \\ de_{ij}^{p} &= 0 \end{aligned} \tag{36}$$

Equation (34) considering principal stress directions gives

$$\frac{de_{11}^{p} - de_{22}^{p}}{T_{11} - T_{22}} = \frac{de_{22}^{p} - de_{33}^{p}}{T_{22} - T_{33}} = \frac{de_{33}^{p} - de_{11}^{p}}{T_{33} - T_{11}} = d\lambda \tag{37}$$

Equation (37) states the Mohar circles of stress and plastic strain increment are similar. Equation (34) can be rewritten in terms of normal stresses giving rise to equations of the form:

$$de_{11}^{p} = \frac{2}{3} d\lambda \left[T_{11} - \frac{1}{2}(T_{22} - T_{33}) \right] \tag{38}$$

Equation (35) thus consists of three equations of the type

$$de_{11} = \frac{2}{3} d\lambda \left[T_{11} - \frac{1}{2}(T_{22} + T_{33}) \right] + \frac{dT_{11} - v(dT_{22} + dT_{33})}{E}$$

and three of the type

$$de_{23} = T_{23}\, d\lambda + \frac{dT_{23}}{2G} \tag{39}$$

Finally, on examining equation (35), it will be seen that the volumetric and deviatoric strain increments can be separated in the expression for the total strain increment. Including the Mises yield criterion, the Prandtl-Ruess equations may then be written as

$$\begin{aligned} de'_{ij} &= S_{ij} d\lambda + \frac{dS_{ij}}{2G} \\ de_{ij} &= \frac{(1-2v)}{E} dT_{ii} \\ S_{ij}S_{ij} &= 2k^2 \end{aligned} \tag{40}$$

These equations for an elastic-plastic solid are usually difficult to handle in a real problem and, in consequence, there are relatively few situations in which these have been employed.

In the case of problems of large plastic flow, the elastic strains are often neglected altogether. The material is then considered as being a perfectly plastic solid. When the stresses are below the yield point, no straining takes place, and the total strain increments are identical. Stresses-strain relations for such type of material were proposed by Levy and von-Mises. By presenting the relations between stress and strain, we have not followed the historical developments of the field. At the present time, it seems more logical to consider the Levy-Mises equations as a separate from of the Prandtl-Ruess equations. The general relationship between strain increment and the reduced stresses was first introduced independently by von-Mises [69]. These equations are now known as Levy-Mises equations and may be written as:

$$\frac{de_{11}}{S_{11}} = \frac{de_{22}}{S_{22}} = \frac{de_{33}}{S_{33}} = \frac{de_{12}}{T_{12}} = \frac{de_{23}}{T_{23}} = \frac{de_{31}}{S_{31}} = d\lambda \tag{41}$$

The superscript p of Eq. (34) may be dropped, since the total Strain increment and plastic strain increment are identical. Further, the Mohar circles of stress and strain increment are identical. In terms of total stresses, the Levy-Mises relation has three equations of the type,

$$de_{11} = \frac{2}{3}d\lambda\left[T_{11} - \frac{1}{2}\left(T_{22} + T_{33}\right)\right]$$

and three of the type

$$de_{23} = T_{23}d\lambda \tag{42}$$

Such that the elastic strain is not taken into account, The Levy-Mises relations obviously cannot be used to obtain information about 'Elastic Spring-back' or residual stresses. For the more complex situations, Prandtl-Ruess equation must be used.

8.6.3 CREEP STATE

Most of creep calculations are on engineering structures. The creep conditions are complex. Laboratory creep tests with complex stress conditions present technical difficulties and the experiments must be performed very carefully if the results are to be reasonably reliable. Therefore, the available experimental data is slender and does not provide a reliable basis for a creep theory that is capable of describing the behavior of material under complex stresses. Moreover, tests can only be made with plane stresses and we have no information about creep performance with stresses along the three axes.

Like plasticity theory, the theory of creep under complex stress is based on certain speculative considerations, which are only partially confirmed experimentally. The simplest case of creep is that in which the great majority of the experimental results relate. Creep of this kind can be described fairly simply as a kind of viscous flow. There are various ways in which the theory can be extended to varying stresses state and number of possibilities is much greater than in the theory of plasticity. In real objects, the nature of stressed state usually varies comparatively little with time, and therefore, the different theories lead to results which are very different from one another.

For steady state of creep, Odqvist [39] has formulated the constitutive equations by considering the rate of strain energy function $\dot{W}$ with von-Mises yield criterion. It relates the rate of steady state of creep to the second invariant of the stress deviator tensor in the following form:

$$e_{ij} = \frac{\partial \dot{W}}{\partial S_{ij}} = \frac{3}{2}\left(\frac{\sigma_e}{\sigma_c}\right)^{n-1} \frac{S_{ij}}{\sigma_c} \tag{43}$$

where e_{ij}, S_{ij}, and σ_c are the strain tensor, stress deviator tensor and effective stress respectively and σ_c and n are material constants.

Stress-strain relations in this from are mostly used to analyze the creep problems based on the following hypothesis:

- Material is incompressible;
- Creep rate is independent of superimposed hydrostatic pressure;
- Existence of flow potential with von-Mises yields condition;
- Material is isotropic;
- Norton's law holds in the special case, i.e., for uniaxial case.

An alternate approach to the problem of multi-axial stationary creep was made possible by Wahl [62]. It consists in using the maximum shear stress (Tresca) as a stress invariant together with the associated flow rule for the body relations. It has already been pointed in section 1.4 that transition state exists when a material goes from elastic state to plastic and then to creep state. In classical treatment, different constitutive equations are used for each state, which are based on some hypothesis that simplifies the problem to a large extent. Firstly, the deformations are assumed to be small to make infinitesimal strain theory applicable. Secondly, the constitutive equations of the material are simplified by assuming incompressibility of the material. In some cases without this assumption, it is not even possible to find the solution of the problem in closed form. By using Seth's transition theory, it has been shown [11–27, 32, 45–48] that the same constitutive equations can be used for different states, though the elastic constants have different meanings in each state.

Gupta and Sharma [26] investigated creep stresses and strain rates in a thin rotating disc having variable thickness and variable density by using Seth's transition theory. It has been concluded that a rotating disc whose density and thickness ratio decrease radially is on the safer side of the design in comparison to a flat disc having variable density.

Singh and Ray [64] operated on modeling the anisotropy and creep in orthotropic Al-SiC composite rotating disc. Analysis of steady state creep in a rotating disc made of composites containing SiC_w has been carried out using Hill yield criterion. The results acquired have been equated with the results attained using von Mises yield criterion for the isotropic composites. The material parameters characterizing anisotropy have been determined from yield stresses. It is perceived that tangential stresses distribution is lower in the middle of the disc but higher near the inner and outer radius but the radius stress distribution does not get significantly affected due to anisotropy.

Ahmet N. Eraslan, Yusuf Orcan [1] have established analytical model based on Tresca's yield criterion and its associated flow rule is developed to analyze the thermo-elasto-plastic response of a linearly hardening cylinder subjected to a non-uniform heat source and convective heat transfer condition at the external boundary. Closed form solutions are obtained for a state of generalized plane strain in different stages of elastic-plastic deformation. In view of a cylinder that is constrained axially, the plastic deformation commences at the axis and depending on the convective boundary condition, another plastic region may develop at the surface before the fully plastic state is reached. A calculation procedure is developed to determine the critical values of the heat-transfer coefficient, for which a third plastic region at the surface does not occur. The effect of various parameters on the critical heat-transfer coefficient is investigated. The residual stresses attained upon unloading are also determined and it is shown that the primary and secondary stresses fall within the shakedown regime.

Hulsurkar [34] applied Seth's transition theory of elastic-plastic and creep deformations to solve the problem of creep in composite cylinders subjected to uniform internal pressure. The generalized expressions for creep transition stresses were obtained, which, in a special case reduce to those derived by assuming the creep laws.

Sharma and co-workers [51, 52, 62] worked on various applications of Seth transition theory to calculate transition elastic plastic and creep stresses and strains in rotation cylinders. Transversely isotropic thick walled cylinders were considered subjected to high rotational forces to analyze the state of stress. Another application of thick walled cylinder subjected to internal pressure was also considered. The authors have tried to explore the impact of non homogeneity of materials on the transitional stress distributions.

Pankaj Thakur and co-workers [42, 43, 44] have comprehensively premeditated creep in the annular disc using Seth' transition theory under different conditions like disc with variable thickness, thin rotating disc with shaft at different temperature, fully plastic thin rotating disc, thin rotating disc having variable density with inclusion, rotating disc with inclusion and made of compressible material with different angular speeds, etc.

Pankaj, Satya Bir Singh, Jasmina Lozanović Šajić [41] functioned on a comprehensive exploration of the influence of non-uniform heat generation subjected to pressure. Effect of heat increased the values of stress for compressible material at the inner surface. Thermal creep stresses and strain rates in a circular disc with a shaft having variable density were determined using Seth transition theory. A similar criterion was used to evaluate creep stresses and strain rate in a rotating disc with respect to changes in mechanical load and thickness profile. Seth's theory of transition has been used in this study. It has been observed that stresses increases with increase in mechanical load and maximum value of strain rate further increases at the internal surface for compressible materials. It is concluded that, rotating disc is likely to fracture by cleavage close to the shaft at the bore.

Tania Bose and Minto Rattan [66] presented the steady state creep behavior of isotropic rotating disc made of parabolically varying functionally graded material in the presence of thermal gradient. The creep rates have been obtained for the discs rotating at elevated temperatures. Investigations for disc rotating at uniform temperature from inner to outer radii has been done using von Mises yield criterion. Further, work has been extended for discs rotating at parabolically decreasing temperature. The results are exhibited graphically for the said temperature profiles. A small variation is observed for radial and tangential stresses for said thermal gradations. However, in the presence of thermal gradation the strain rates vary significantly as compared to disc at uniform temperature. Thus in functionally graded rotating disc the temperature gradation significantly affects the creep behavior of a rotating disc.

Shivdev Shahi et al. [63] studied creep parameter in disc made of transversely isotropic and isotropic material subjected to shaft by using Seth's transition theory. Neither the yield criterion nor the associated flow rule is assumed here. The results obtained here are applicable to transversely isotropic and isotropic materials. If the additional condition of incompressibility is imposed, then the expression for stresses corresponds to those arising from Tresca yield condition. It has been observed that radial stress has maximum value at the inner surface of the rotating disc made of isotropic material as compared to the circumferential stress and this value of radial stress further increases with the increase in angular speed. Strain rates have maximum values at the inner surface for transversely isotropic material.

Equations of equilibrium for a body having variable thickness h in the radial direction, in cylindrical polar coordinates are given by [58]

$$\frac{\partial}{\partial r}(hT_{rr})+\frac{h}{r}\left(\frac{\partial T_{r\theta}}{\partial\theta}\right)+h\left(\frac{\partial T_{rz}}{\partial z}\right)+\frac{h}{r}(T_{rr}-T_{\theta\theta})+hF_r=0$$
$$\frac{\partial}{\partial r}(hT_{r\theta})+\frac{h}{r}\left(\frac{\partial T_{\theta\theta}}{\partial\theta}\right)+h\left(\frac{\partial T_{\theta z}}{\partial z}\right)+2\frac{h}{r}T_{r\theta}+hF_\theta=0$$
$$\frac{\partial}{\partial r}(hT_{zr})+\frac{h}{r}\left(\frac{\partial T_{\theta z}}{\partial\theta}\right)+h\left(\frac{\partial T_{zz}}{\partial z}\right)+2\frac{h}{r}T_{rz}+hF_z=0 \qquad (44)$$

where F_r, F_θ F_z are the body forces along r, θ, z directions. Also

$$h=h_0\left(\frac{r}{b}\right)^{-k} \qquad (45)$$

where, h_0, k is positive real constants and b is the external radius.

Following Ref. [60] for a body having constant thickness, Eq. (41) become

$$\frac{\partial}{\partial r}(T_{rr})+\frac{1}{r}\frac{\partial T_{r\theta}}{\partial\theta}+\frac{\partial T_{rz}}{\partial z}+\frac{1}{r}(T_{rr}-T_{\theta\theta})+F_r=0,$$
$$\frac{\partial}{\partial r}(T_{r\theta})+\frac{1}{r}\frac{\partial T_{\theta\theta}}{\partial\theta}+\frac{\partial T_{\theta z}}{\partial z}+\frac{2}{r}T_{r\theta}+F_\theta=0,$$
$$\frac{\partial}{\partial r}(T_{zr})+\frac{1}{r}\frac{\partial T_{\theta z}}{\partial\theta}+\frac{\partial T_{zz}}{\partial z}+\frac{2}{r}T_{rz}+F_z=0 \qquad (46)$$

For a rotating disc with the axis of rotation set as z axis, the body force is the centrifugal force with components

$$F_r = \rho\omega^2 r; F_\theta = 0; F_z = 0$$

where ρ is the density of the material of the disc.

In the present case, we have an axisymmetrical body under the action of axisymmetrical force. Hence, the stress components τ_{rr} and $\tau_{\theta\theta}$ are functions of r only and

$$\tau_{r\theta} = \tau_{\theta z} = \tau_{zr} = \tau_{zz} = 0$$

KEYWORDS

- **creep**
- **elastic**
- **plastic**
- **strain**
- **stress**
- **transition**

REFERENCES

1. Ahmet N, Eraslan and Yusuf Orcan (2004). A parametric analysis of rotating variable thickness elastoplastic annular disks subjected to pressurized and radially constrained boundary conditions, *Turkish J. Eng. Env. Sci., 28,* 381–395.
2. Biot, M. A., (1939). Non-linear theory of elasticity and the linearized case for a body under initial stress. *Phil Magazine, 27*, 449–452, 468–469.
3. Boyle, J. T., & Spence, J., (1983). *Stress Analysis or Creep*. Butterworths Co. Ltd. London.
4. Chadravotry, J., (1987). *Theory of Plasticity*. McGraw-Hill Book Coy, New York.
5. Chung, D. T., Horgam, C. O., & Abeyarathne, R., (1986). Finite deformation of internally pressurized hollow cylinders and spheres. *Int. J. Solid Structure, 22*, 1557–1570.
6. Cossert, E. F., (1896). Surla theorie, de 1, 'Elasticite, annals de I' Universite' de Toulouse Pour. *Mathematiqueet Physique, 10*.
7. Dorn, J. E., (1961). *Mechanical Behavior of Materials at Elevated Temperatures.* McGraw-Hill, New York.
8. Durban, D., (1987). An exact solution for the internally pressurized elastic-plastic strain hardening annular plate. *Acta. Mechanica., 66*, 111.

9. Fung, Y. C., (1965). *Foundation of Solid Mechanics.* Englewood Cliffs, NJ, Prentice-Hall.
10. Gamer, U., (1984). The elastic-plastic stress distribution In: *A Rotating Annulus and in the Annulus Under External Pressure, ZAMM* (Vol. 64, pp. 126–128).
11. Gamer, U., (1983). Trescas yield condition and the rotating disc. *Journal of Applied Mechanics, ASME Trans., 50*, 676–678.
12. Goodier, J. N., & Hodge, P. G. Jr., (1960). *Structural Mechanics.* Pergamon Press, Oxford.
13. Güaven, U., (1992). Elastic-plastic stress on a rotating annular disc of variable thickness and variable density. *Inst. Mech., Sci., 34*(2), 133–138.
14. Güaven, U., (1991). Elastic-plastic stress distribution in the rotating disc with variable thicknesses. *Archive of Appl. Mech., 61*, 549–554.
15. Güaven, U., (1993). On the Stresses in an elastic-plastic annular disc of variable thickness under external pressure. *Int. J. Solids Structure, 30*(5), 651–658.
16. Gupta S. K., & Shukla R. K., (1993). PhD Thesis, Department of Mathematics (pp. 43–53). H.PU. Shimla, India.
17. Gupta, S. K., & Dharmani, R. L., (1981). Creep transition in rotating cylinders. *J. Math. Phy. Sci., 15*(6), 525–536.
18. Gupta, S. K., & Pankaj, (2007). Creep transition in thin rotating disc with rigid inclusion. *Defense Science Journal, 57*, 185–195.
19. Gupta, S. K., & Pankaj, (2008). Creep Transition in an isotropic disc having variable thickness subjected to internal pressure. *Proc. Nat. Acad. Sci. India Sect. A., 78*, Pt. I.
20. Gupta, S. K., & Pankaj, (2007). Thermo elastic-plastic transition in a thin rotating disc with inclusion. *Int. Scientific J. Thermal Sci., 11*(1), 103–118.
21. Gupta, S. K., & Pathak, S., (2000). Elastic-plastic transition in a thin rotating disc having variable density with edge load. *Proc. Nat. Acad. Sci., India, 70*(a), I, 75–86.
22. Gupta, S. K., & Pathak, S., (2000). Creep transition in thin rotating disc of variable density. *Defense Science J., 50*(2), 147–153.
23. Gupta, S. K., & Pathak, S., (2001). Thermo creep transition in a thick-walled circular cylinder under internal pressure. *Indian J. of Pure and Appl. Math., 32*(2), 237–253.
24. Gupta, S. K., & Rana, V. D., (1983). Thermo elastic-plastic transition in rotating cylinders. *Indian Jr. Tech., 21*, 499–502.
25. Gupta, S. K., & Rana, V. D., (1989). Thermo elastic-plastic and creep transition in rotating cylinders. *J. Math. Phy. Sci., 23*, 71–90.
26. Gupta, S. K., & Sharma, S., (2000). Creep Transition in a thin rotating disc having variable thickness and variable density. *Indian Journal of Pure and Appl. Math., 31*(10), 1235–1248.
27. Gupta, S. K., & Sharma, S., (1998). Thermo creep transition of non-homogeneous thick-walled circular cylinder under internal pressure. *Indian Journal of Pure and Applied Mathematics, 29*(11), 1111–1125.
28. Gupta, S. K., & Shukla, R. K., (1994). Elastic-plastic transition in a thin rotating disc. *Ganita, 45*, 79–85.
29. Gupta, S. K., (1985). Elastic-plastic and creep transition in rotating cylinder. *Proceedings of the Workshop on Solid Mechanics* (pp. 153–162). March 13th–16th University of Roorkee, Roorkee.

30. Gupta, S. K., (1989). Thermal creep transition in an isotropic rotating cylinder. *Jr. Math. and Phy. Sci., 23*(2), 147–159.
31. Gupta, S. K., Sharma, S., & Pathak, S., (1998). Elastic-plastic transition in thin rotating disc of variable thickness with edge load. *Ganita, 49*(1), 61–75.
32. Gupta, S. K., (1980). Elastic-plastic and creep transition of thick-walled cylinder under uniform pressure. *Proc. of Int. Symp. on Non-Linear Mech.* (pp. 169–175). Kharagpur.
33. Hill, R., (1980). *The Mathematical Theory of Plasticity.* Oxford Univ. Press, London.
34. Hulsurkar, S., (1979). On transition function in a cylinder under uniform pressure. *J. Math. Phy. Sci., 13*(3), 209–217.
35. Jonson, W., & Mellor, P. B., (1971). *Plasticity for Mechanical Engineers.* Van Nostrand Reinhold Company, London.
36. Kraus, H., (1980). *Creep Analysis* (pp. 568–599). John Wiley & Sons, New York: Toronto.
37. Lubhan, D., & Felger, R. P., (1961). *Plasticity and Creep of Metals*. Wiley, New York.
38. Nabarro, F. R. N., & Villiers, H. L. D., (1995). *Physics of Creep.* Taylor & Francis, P. A.
39. Nadai, A., (1950). *Theory of Flow and Fracture of Solids.* McGraw- Hill Book Coy, Inc., New York.
40. Odqvist, F. K. G., (1959). Engineering theories of creep. *Proc. of the Vth Congress on Theo. Appl. Mech., 81*.
41. Pankaj Thakur, Singh, S. B., & Sajic, J. L., (2015), "Thermo Elastic-Plastic Deformation in a Solid Disk with Heat Generation Subjected to Pressure," *Structural Integrity and Life, 15*(3), 135–142.
42. Pankaj, T., Sethi, M., Shivdev, S., Singh, S. B., & Emmanuel, F. S., (2018). Exact solution of rotating disc with shaft problem in the elastoplastic state of stress having variable density and thickness. *Structural Integrity and Life, 18*(2), 126–132.
43. Pankaj, T., Sethi, M., Shivdev, S. I., Singh, S. B., Emmanuel, F. S., & Lozanović, Š. J., (2018). Modeling of creep behavior of a rotating disc in the presence of load and variable thickness by using Seth transition theory. *Structural Integrity and Life, 18*(2), 133–140.
44. Pankaj, T., Shivdev, S., Nishi, G., & Singh, S. B., (2017). Effect of mechanical load and thickness profile on creep in a rotating disc by using Seths transition theory. *American Institute of Physics: Conference Proceedings (USA), 1859*, 20–24.
45. Parkus, H., (1976). *Thermo-Elasticity, Springer-Verlag Wien*. New York, USA.
46. Penny, R. K., & Mariott, D. L., (1995). *Design for Creep.* Chapman and Hall, London.
47. Ritz, P. M., (1948). Large deformation and plasticity. *Academia Nauk. SSR Prikladnaya Mathematika I Mechniks, 12,* 211–212.
48. Rivilin, R. S., (1948). Large elastic deformation of isotropic materials. *I, II, III, IV Philos. Trans. Roy. Soc. London, 240*, 459–525.
49. Robotnov, Y. N., (1969). In: Leckie, F. A., (ed.), *Creep Problems in Structural Members.* Translated from Russian Edition, Moscow. North-Holland, Amsterdam.
50. Robotnov, Y. N., (1966). Kinetics of creep and creep rupture. *Proc. IUTAM Symposia on Irreversibility and Transfer of Physics Characteristics in a Continuum* (pp. 325–334). Vienna.

51. Sharma, S., & Sahni, M., (2008). Creep transition of transversely isotropic thick-walled rotating cylinder. *Adv. Theory. Appl. Mech., 1*(7), 315–325.
52. Sharma, S., (2004). Elastic-plastic transition of non-homogeneous thick-walled circular cylinder under internal pressure. *Def. Sc. Journal, 54*(2).
53. Seth, B. R., (1972). Yield conditions in plasticity. *Arch. Mech., Stos., 24*(5/6), 769–776.
54. Seth, B. R., (1959). An application of theory of finite strain. *Proc. Indian Acad. Sci., (A), 9*, 17–19.
55. Seth, B. R., (1972). Creep transition. *J. Math. Phys. Sci., 6*, 1.
56. Seth, B. R., (1966). Measure concept in mechanics. *Int. J. Non-Linear Mech., I*, 35–40.
57. Seth, B. R., (1970). Transition analysis of collapse of thick-walled cylinders. *ZAMM, 50*, 617–621.
58. Seth, B. R., (1962). Transition theory of elastic-plastic deformation, creep, and relaxation. *Nature, 195*, 896–897.
59. Seth, B. R., (1970). Creep rupture, IUTAM Symp.: On creep in structures. *Gothenberg*, 167–169.
60. Seth, B. R., (1974). Creep transition in rotating cylinder. *J. Math. Phys. Sci., 8*(1), 1–5.
61. Seth, B. R., (1963). Elastic-plastic transition in shells and tubes under pressure. *ZAMM, 43*, 345.
62. Sharma, S., (1999). Some problems in elastic-plastic and creep transition for non-homogeneous materials. PhD Thesis, Department of Maths, H.P.U., Shimla.
63. Shivdev, S., Singh, S. B., & Pankaj, T., (2019). Modeling creep parameter in rotating discs with rigid shaft exhibiting transversely isotropic and isotropic material behavior. *The Journal of Emerging Technologies and Innovative Research, 6*(1), 387–395.
64. Singh, S. B. & Ray, S. (2001). "Steady-state creep behavior in an isotropic functionally graded material rotating disc of Al-SiC composite," *Metallurgical Transactions 32A*(7), 1679–1685.
65. Sokolnikoff, I. S., (1950). *Mathematical Theory of Elasticity* (pp. 29–31, 56–58). McGraw-Hill Pub. Company, New York.
66. Tania Bose & Rattan, M. (2018). "Modeling Creep Analysis of Thermally Graded Anisotropic Rotating Composite Disc." *International Journal of Applied Mechanics 10*(6), 1850063.
67. Timoshenko, S. P., & Goodier, J. N., (1951). *Theory of Elasticity* (3rd edn.). McGraw-Hill Book Coy, New York, London.
68. Tresca, H., (1868). 'M' emoire Sur I' Ecoulement Descorps Solids. *M Emories Presents Par Divers Savents, 81*, 733–799.
69. Olszak, W., & Urbanowski, W., (1955). Elastic-plastic thick-walled non-homogeneous cylinder subjected to internal pressure and longitudinal load. *AMS, 7*(3), 315–336.
70. Wahl, A. M., (1956). Analysis of creep in rotating discs based on Tresca criterion and associated flow rule. *Jr. Appl. Mech., 23*, 103.
71. Zhilun, X. U., (1993). *Applied Elasticity.* Wiley Eastern Limited.
72. Zvolinsky, N. (1939). On certain problems of non-linear theory of elasticity. *Akad. Nauk. Lbid., 2*(4).

CHAPTER 9

COMPARATIVE ANALYSIS OF ELASTIC-PLASTIC STRESS DISTRIBUTIONS IN HUMAN FEMUR BONE, TITANIUM, AND BORON-ALUMINUM FIBER-REINFORCED COMPOSITE FOR SURGICAL IMPLANTS AND PROSTHETIC EQUIPMENT DESIGN

SHIVDEV SHAHI,[1] SATYA BIR SINGH,[1] and PANKAJ THAKUR[2]

[1]*Department of Mathematics, Punjabi University Patiala, Punjab – 147002, India, E-mails: shivdevshahi93@gmail.com (Shivdev Shahi), sbsingh69@yahoo.com (S. B. Singh)*

[2]*Department of Mathematics, ICFAI University, Himachal Pradesh – 174103, India*

ABSTRACT

In this chapter, elastic-plastic stress distributions in human femur bone are calculated analytically. The bone is modeled in the form of a cylinder which exhibits orthotropic macroscopic symmetry. Such behavior is characterized typically for unidirectional laminae with on axis loading. Seth's transition theory has been used to model the elastic-plastic state of stresses. The cylinder so modeled is subjected to external pressure. The results obtained are compared to cylinders made of boron-aluminum fiber reinforced composite which also exhibits orthotropic material behavior and Titanium which exhibits transversely isotropic material behavior. The results obtained infer to the fact that boron-aluminum fiber reinforced composite has a much closer resemblance with the elastic plastic behavior of bone structures and will be a better option as a base material for Surgical implants and prosthetic equipment design as compared to Titanium.

9.1 INTRODUCTION

Bones are both anisotropic and heterogeneous in their mechanical properties. Humans have evolved in a manner to bear an erect posture. The posture is supported by the bones and muscles of hind limbs which carry a maximum weight of the body. The limbs are primarily comprised of the femur and the tibia bone along with smaller bones of knee, ankle, and foot. Due to various forces to which the bone is subjected, there is a continuous state of stress which not only depends on the intensity and the manner of application of force but also on the mechanical properties of the bone structure. Bundy [1] experimentally demonstrated for human femoral bone that the mechanical properties vary along the length of the bone which supports the orthotropic elastic behavior. He also showed that the bone is anisotropic but did not measure all of the elastic constants needed to completely characterize the anisotropy. Most investigators examining the anisotropy of bone have assumed it to be transversely isotropic and have measured the five elastic constants needed to characterize such a material [8, 21]. None of these investigators appears to have systematically examined the heterogeneous nature of these properties. The investigation of stress distribution is of prime significance because these results are of much help to orthopedic surgeons and the agencies which work on the design of prosthetic limbs. The stress distribution in the bone implants if available at hand, may help to improve the success rates of surgeries. Pitkin et al. [7] proposed to use titanium as a base material to design the pylon of prosthetic legs. Medical grade Titanium is also used inserted in femur and tibia bones in cases of multiple fractures and in cases when the bones are damaged due to osteoporosis [3, 9]. The applications are shown in Figures 9.1 and 9.2. In spite of such compatibilities, titanium has a far lesser tensile strength when compared to composite material made of an aluminum matrix reinforced with boron fibers. Aluminum-based composite is also characterized with a lesser weight as compared to titanium and has a greater stress-bearing capacity. Such applications of aluminum matrix composites and functionally graded materials have been studied by Singh and Ray [14].

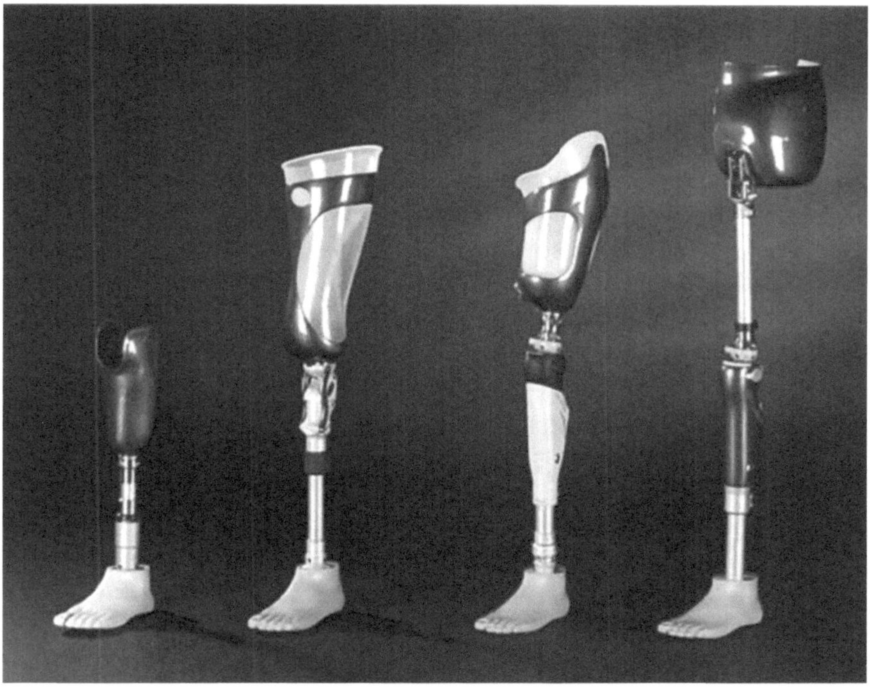

FIGURE 9.1 Cylindrical pylons in prosthetic legs.

Ma [6] investigated the stresses in rotating solid discs having variable thickness and operating under a temperature gradient in the radial direction of the disc using the analysis based on Tresca's criterion [16]. Bhatnagar et al. [2] studied creep behavior of orthotropic rotating discs having variable thickness (constant, linear, and hyperbolic) using Norton's power law. Gupta and Singh [18] further studied the deformation behavior of a functionally graded disc using similar yielding criteria. These works however did not consider the non-linear character of the transition phase which was defined by B. R. Seth [10–12].

Seth's transition theory is used in this chapter to obtain elastic-plastic stresses in human femur bone considering the orthotropic elastic constants obtained using ultrasonic measurement by Buskirk et al. [20]. Similarly, elastic-plastic stresses for composite material made of aluminum reinforced with boron fibers and titanium, modeled in the shape of a closed-ended hollow cylinder subjected to external pressure along the radius, are obtained using Seth's transition theory. Boron-aluminum fiber-reinforced composite, composed of uniaxial boron fibers in a matrix of 6061 aluminum alloy was tested for the stiffness constants using acoustic resonance spectroscopy by Ledbetter et al. [5] has been used along with titanium whose stiffness constants are given by Uyaner et al. [17]. The concept of generalized strain measures and Seth's transition theory [10, 11] has been applied to find elastic-plastic stresses in problems associated to various structural components [13, 15] by solving the non-linear differential equations at the transition points. All these problems based on the recognition of the transition state as a separate state necessitates showing the existence of the constitutive equation for that state.

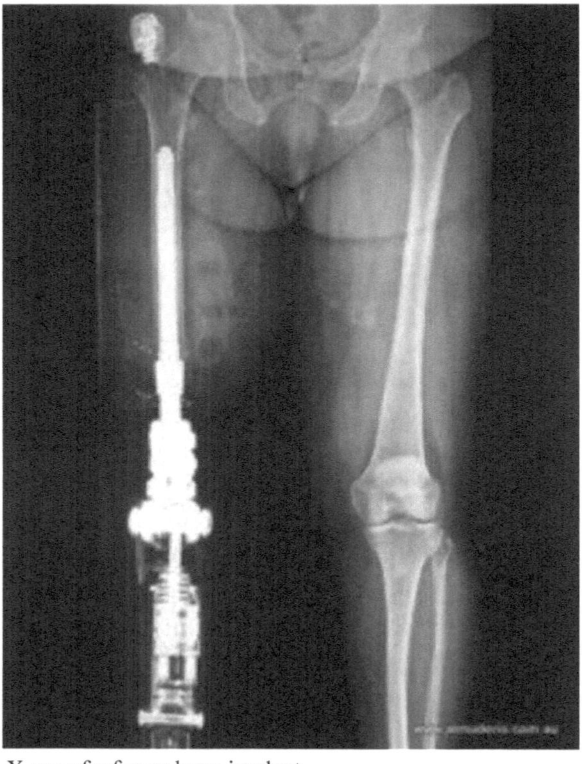

FIGURE 9.2 X-ray of a femur bone implant.

9.2 GOVERNING EQUATIONS

Consider a hollow circular cylinder of internal and external radii a and b, respectively, subject to pressure p_0 as represented in Figure 9.3. The circumferential, axial, and radial stresses in the cylindrical structure will propagate in the directions as shown in Figure 9.4.

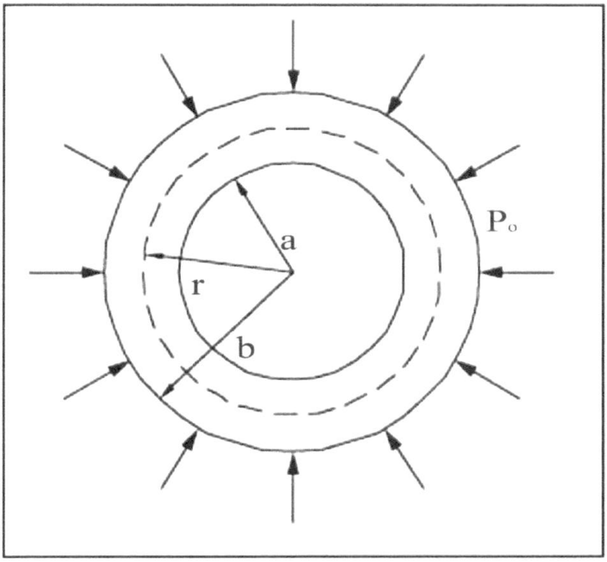

FIGURE 9.3 Geometrical cross-section of cylinder under external pressure.

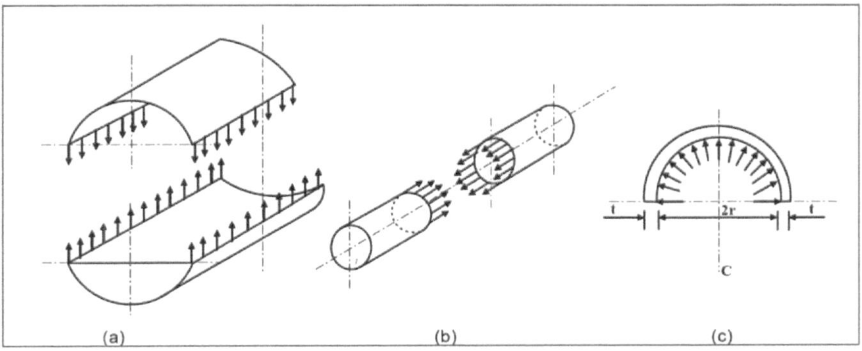

FIGURE 9.4 Geometrical representation [(a) Circumferential stresses (b) Axial stresses (c) Radial stresses].

1. **Displacement Coordinates:** The components of displacement in cylindrical coordinates are taken as:

$$u = r(1-\beta);\; v = 0;\; w = dz \tag{1}$$

where β is position function depending on $r = \sqrt{x^2 + y^2}$ only and d is a constant.

The generalized components of strain are given as:

$$e_{rr} = \frac{1}{2}\left[1-(r\beta'+\beta)^2\right],\, e_{\theta\theta} = \frac{1}{2}\left[1-\beta^2\right],\, e_{zz} = \frac{1}{2}\left[1-(1-d)^2\right],\, e_{r\theta} = e_{\theta z} = e_{zr} = 0. \tag{2}$$

and

$$\beta' = d\beta/dr$$

2. **Stress-Strain Relation:** It is for isotropic material is given by (1):

$$T_{ij} = c_{ijkl}e_{kl},\, (i, j, k, l = 1, 2, 3)$$

where T_{ij} and e_{kl} are the stress and strain tensors respectively. These nine equations contain a total of 81 coefficients e_{ijkl}, but not all the coefficients are independent. The symmetry of T_{ij} and e_{ij} reduces the number of independent coefficients to 36. For elastic orthotropic, materials which have three mutually orthogonal planes of elastic symmetry, these independent coefficients reduce to 12 and to 9 if the coefficients are symmetric. The constitutive equations for orthotropic media are given by Altenbach et al. [1]:

$$\begin{bmatrix} T_{11} \\ T_{22} \\ T_{33} \\ T_{23} \\ T_{31} \\ T_{12} \end{bmatrix} = \begin{bmatrix} c_{11} & c_{12} & c_{13} & 0 & 0 & 0 \\ c_{21} & c_{22} & c_{23} & 0 & 0 & 0 \\ c_{31} & c_{32} & c_{33} & 0 & 0 & 0 \\ 0 & 0 & 0 & c_{44} & 0 & 0 \\ 0 & 0 & 0 & 0 & c_{55} & 0 \\ 0 & 0 & 0 & 0 & 0 & c_{66} \end{bmatrix} \begin{bmatrix} e_{11} \\ e_{22} \\ e_{33} \\ e_{23} \\ e_{31} \\ e_{12} \end{bmatrix}, \tag{3}$$

Substituting Eq. (2) in Eq. (3), we get:

$$T_{rr} = \frac{c_{11}}{2}\left[1-(r\beta'+\beta)^2\right]+\frac{c_{12}}{2}\left(1-\beta^2\right)+\frac{c_{13}}{2}\left(1-(1-d)^2\right);$$
$$T_{\theta\theta} = \frac{c_{21}}{2}\left[1-(r\beta'+\beta)^2\right]+\frac{c_{22}}{2}\left(1-\beta^2\right)+\frac{c_{23}}{2}\left(1-(1-d)^2\right);$$
$$T_{zz} = \frac{c_{31}}{2}\left[1-(r\beta'+\beta)^2\right]+\frac{c_{32}}{2}\left(1-\beta^2\right)+\frac{c_{33}}{2}\left(1-(1-d)^2\right);$$
$$T_{r\theta} = T_{\theta z} = T_{zr} = 0 \tag{4}$$

3. **Equation of Equilibrium:** The equations of equilibrium are all satisfied except:

$$\frac{d}{dr}T_{rr} + \frac{T_{rr}-T_{\theta\theta}}{r} = 0 \tag{5}$$

4. **Critical Points or Turning Points:** By substituting Eq. (4) into Eq. (5), we get a non-linear differential equation with respect to β:

$$\beta\frac{dP}{d\beta} = \frac{[(c_{11}-c_{21})\left\{1-\beta^n(P+1)^n\right\}+(c_{12}+c_{22})\left(1-\beta^n\right)+(c_{13}-c_{23})\{1-(1-d)^n\}]}{nc_{11}\beta^n P(P+1)^{n-1}} - \frac{c_{12}\left(1-\beta^n\right)}{c_{11}(P+1)^{n-1}} - (P+1) \tag{6}$$

where P is function of β and β is function of r only.

5. **Transition Points:** The transition points of β in Eq. (6) are $P \to 0, P \to -1$ and $P \to \pm\infty$.

   The only critical point of interest $P \to \pm\infty$ which is sufficient for the elastic-plastic state given by Seth [10, 11].

6. **Boundary Condition:** The boundary conditions of the problem are given by:

$$r = a,\ \tau_{rr} = 0$$
$$r = b,\ \tau_{rr} = -p_o \tag{7}$$

9.3 PROBLEM SOLUTION

For finding the elastic-plastic stress, the transition function is taken through the principal stress [13, 15] at the transition point $P \to \pm\infty$, we define the transition function ζ as:

$$\zeta = 1 - \frac{nT_{rr}}{\left(c_{11} + c_{12} + c_{13}\right)} \cong \frac{1}{\left(c_{11} + c_{12} + c_{13}\right)}\left[c_{11}\beta^{n}\left(P+1\right)^{n} + c_{12}\beta^{n} + c_{13}(1-d)^{n}\right] \quad (8)$$

where ζ be the transition function of r only.

Substituting the value of T_{rr} from Eq. (4) in Eq. (8)

$$\zeta = \frac{c_{11}}{\left(c_{11} + c_{12} + c_{13}\right)}\left[\beta^{n}\left(P+1\right)^{n} + \frac{c_{12}}{c_{11}}\beta^{n} + \frac{c_{13}}{c_{11}}(1-d)^{n}\right] \quad (9)$$

Taking the logarithmic differentiation of Eq. (9), with respect to r and using Eq. (6), we get:

$$\frac{d\left(\log\zeta\right)}{dr} = \left(\frac{n\beta^{n}P\left[\left(P+1\right)^{n} + \beta\frac{dP}{d\beta}\left(P+1\right)^{n-1} + \frac{c_{12}}{c_{11}}\right]}{r\left[\beta^{n}\left(P+1\right)^{n} + \beta^{n}\frac{c_{12}}{c_{11}} + \frac{c_{13}}{c_{11}}(1-d)^{n}\right]}\right) \quad (10)$$

Taking the asymptotic value of Eq. (10) as $P \to \pm\infty$ and integrating, we get

$$\zeta = A_{0}r^{-K} \quad (11)$$

where A_0 is a constant of integration and $K = c_{11} - c_{21}/c_{11}$. From Eqs. (8) and (11), we have

$$T_{rr} = \frac{\left(c_{11} + c_{12} + c_{13}\right)}{n}\left[1 - A_{0}r^{-K}\right] \quad (12)$$

The value of material constant in the transition range is given by

$$Y = \frac{1}{n}\frac{\left[c_{11}c_{22}c_{33} - c_{11}c_{23}^{2} - c_{22}c_{12}^{2} - c_{33}c_{12}^{2} + 2c_{23}c_{12}^{2}\right]}{\left[c_{22}c_{33} - c_{23}^{2}\right]} \quad (13)$$

where Y is yield stress in tension.

Using the yield stress in Eq. (12), we get:

$$T_{rr} = \frac{\left(c_{11} + c_{12} + c_{13}\right)Y(c_{22}c_{33} - c_{23}^{2})\left[1 - A_{0}r^{-K}\right]}{\left[c_{11}c_{22}c_{33} - c_{11}c_{23}^{2} - c_{22}c_{12}^{2} - c_{33}c_{12}^{2} + 2c_{23}c_{12}^{2}\right]} \quad (14)$$

Substituting Eq. (14) in equation of equilibrium

$$T_{\theta\theta} - T_{rr} = \frac{(c_{11} + c_{12} + c_{13})Y(c_{22}c_{33} - c_{23}^2)A_0(c_{11} - c_{21})r^{-K}}{c_{11}\left[c_{11}c_{22}c_{33} - c_{11}c_{23}^2 - c_{22}c_{12}^2 - c_{33}c_{12}^2 + 2c_{23}c_{12}^2\right]} \quad (15)$$

$$T_{\theta\theta} = \frac{(c_{11} + c_{12} + c_{13})Y(c_{22}c_{33} - c_{23}^2)\left[1 - (1-K)A_0 r^{-K}\right]}{\left[c_{11}c_{22}c_{33} - c_{11}c_{23}^2 - c_{22}c_{12}^2 - c_{33}c_{12}^2 + 2c_{23}c_{12}^2\right]} \quad (16)$$

Using boundary conditions of Eq. (7) in Eq. (14) we get

$$A_0 = a^K \quad (17)$$

and

$$p_o = \frac{(c_{11} + c_{12} + c_{13}).Y.(c_{22}c_{33} - c_{23}^2)\left[\left(\frac{a}{b}\right)^K - 1\right]}{\left[c_{11}c_{22}c_{33} - c_{11}c_{23}^2 - c_{22}c_{12}^2 - c_{33}c_{12}^2 + 2c_{23}c_{12}^2\right]} \quad (18)$$

Substituting Eq. (17) in the principal stresses and principal stress difference, we get:

$$T_{rr} = \frac{(c_{11} + c_{12} + c_{13})Y(c_{22}c_{33} - c_{23}^2)\left[1 - \left(\frac{a}{r}\right)^K\right]}{\left[c_{11}c_{22}c_{33} - c_{11}c_{23}^2 - c_{22}c_{12}^2 - c_{33}c_{12}^2 + 2c_{23}c_{12}^2\right]} \quad (19)$$

$$T_{\theta\theta} - T_{rr} = \frac{(c_{11} + c_{12} + c_{13})Y(c_{22}c_{33} - c_{23}^2)(c_{11} - c_{21})\left(\frac{a}{r}\right)^K}{c_{11}\left[c_{11}c_{22}c_{33} - c_{11}c_{23}^2 - c_{22}c_{12}^2 - c_{33}c_{12}^2 + 2c_{23}c_{12}^2\right]} \quad (20)$$

$$T_{zz} = \frac{c_{31}(T_{\theta\theta} + T_{rr})}{c_{11} + c_{21}} + \left[c_{32} - \frac{c_{31}(c_{12} + c_{22})}{c_{11} + c_{21}}\right]\frac{1}{n} + \left[c_{33} - \frac{c_{31}(c_{13} + c_{23})}{c_{11} + c_{21}}\right]e_{zz} \quad (21)$$

$$T_{zz} = \frac{c_{31}(T_{\theta\theta} + T_{rr})}{c_{11} + c_{21}} + \left[c_{32} - \frac{c_{31}(c_{12} + c_{22})}{c_{11} + c_{21}} \right] \frac{1}{n} + \left[c_{33} - \frac{c_{31}(c_{13} + c_{23})}{c_{11} + c_{21}} \right] e_{zz} \quad (22)$$

where $e_{zz} = \frac{1}{n}\left[1-(1-d)^n\right]$ and e_{zz} is obtained by considering the cylinder as closed-ended,

$$2\pi \int_a^b rT_{zz}dr = \pi b^2 p$$

and

$$e_{zz} = \frac{\left[\frac{b^2 p}{(b^2 - a^2)} \left(1 - \frac{2c_{31}}{c_{11} + c_{21}} \right) - \left[c_{32} - \frac{c_{31}(c_{12} + c_{22})}{c_{11} + c_{21}} \right] \frac{1}{n} \right]}{\left[c_{33} - \frac{c_{31}(c_{13} + c_{23})}{c_{11} + c_{21}} \right]} \quad (23)$$

Substituting the value of e_{zz} and the values of T_{rr} and $T_{\theta\theta}$ from Eqs. (22) and (23) we get:

$$T_{zz} = \frac{c_{31}}{c_{11} + c_{21}} Y \frac{(c_{11} + c_{12} + c_{13})(c_{22}c_{33} - c_{23}^2)\left[2 - 2\left(\frac{a}{r}\right)^K + K\left(\frac{a}{r}\right)^K \right]}{\left[c_{11}c_{22}c_{33} - c_{11}c_{23}^2 - c_{22}c_{12}^2 - c_{33}c_{12}^2 + 2c_{23}c_{12}^2 \right]} + \frac{b^2 p}{(b^2 - a^2)}\left(1 - \frac{2c_{31}}{c_{11} + c_{21}} \right) \quad (24)$$

1. **Initial Yielding**: $|T_{\theta\theta} - T_{rr}|$ is maximum at r = b which clearly shows that the yielding will take place at the outer surface. Hence, we have

$$|T_{\theta\theta} - T_{rr}| = \frac{(c_{11} + c_{12} + c_{13})Y(c_{22}c_{33} - c_{23}^2)(c_{11} - c_{21})\left(\frac{a}{b}\right)^K}{c_{11}\left[c_{11}c_{22}c_{33} - c_{11}c_{23}^2 - c_{22}c_{12}^2 - c_{33}c_{12}^2 + 2c_{23}c_{12}^2 \right]} \cong Y^* \text{ (Yielding)} \quad (25)$$

Substituting the value of Y in terms of Y* in Eqs. (19), (21), and (24) the transitional stresses are given by

$$T_{rr} = \frac{1}{K}\left(\frac{b}{a}\right)^K Y^* \left[1 - \left(\frac{a}{r}\right)^K \right] \quad (26)$$

$$T_{\theta\theta} = \frac{1}{K}\left(\frac{b}{a}\right)^{K} Y^{*}\left[1-(1-K)\left(\frac{a}{r}\right)^{K}\right] \tag{27}$$

$$T_{zz} = \frac{c_{31}}{c_{11}+c_{21}} Y^{*}\frac{1}{K}\left(\frac{b}{a}\right)^{K}\left[2-2\left(\frac{a}{r}\right)^{K}+K\left(\frac{a}{r}\right)^{K}\right]+\frac{b^{2}p_{i}}{(b^{2}-a^{2})}\left(1-\frac{2c_{31}}{c_{11}+c_{21}}\right) \tag{28}$$

The pressure at initial yielding is calculated using Eqs. (18) and (25)

$$p_{i} = \frac{1}{K}\left(\frac{b}{a}\right)^{K} Y^{*}\left[\left(\frac{a}{b}\right)^{K}-1\right] \tag{29}$$

Converting into nondimensional components

$R = r/a,\ R_0 = b/a,\ \sigma_{rr} = T_{rr}/Y^*,\ \sigma_{\theta\theta} = T_{\theta\theta}/Y^*, \sigma_{zz} = T_{zz}/Y^*$ and $P_i = p_i/Y^*$

$$\sigma_{rr} = \frac{1}{K} R_0^{K}\left[1-R^{-K}\right] \tag{30}$$

$$\sigma_{\theta\theta} = \frac{1}{K} R_0^{K}\left[1-(1-K)R^{-K}\right] \tag{31}$$

$$\sigma_{zz} = \frac{c_{31}}{c_{11}+c_{21}}\frac{1}{K} R_0^{K}\left[2-2R^{-K}+KR^{-K}\right]+\frac{p_i}{(1-R_0^{-2})}\left(1-\frac{2c_{31}}{c_{11}+c_{21}}\right) \tag{32}$$

$$P_i = \frac{1}{K} R_0^{K}\left[R_0^{-K}-1\right] \tag{33}$$

2. **Fully Plastic State**: For fully-plastic case, $c_{11} = c_{13} = -c_{12}$, $c_{23} = c_{21} = -c_{22}$ stresses and pressure are calculated as follows [12],

where $R = r/b,\ R_0 = b/a,\ \sigma_{rr} = T_{rr}/Y^*,\ \sigma_{\theta\theta} = T_{\theta\theta}/Y^*,\ K_1 = (c_{11}-c_{22})/c_{11}$

and $P_f = p_f/Y^*$.

$$\sigma_{rr} = \frac{1}{K_1} R_0^{K_1}\left[1-R^{-K_1}\right] \tag{34}$$

$$\sigma_{\theta\theta} = \frac{1}{K_1} R_0^{K_1} \left[1 - (1 - K_1) R^{-K_1}\right] \tag{35}$$

$$\sigma_{zz} = \frac{c_{33}}{c_{11} + c_{21}} \frac{1}{K_1} R_0^{K_1} \left[2 - 2R^{-K_1} + K_1 R^{-K_1}\right] + \frac{p_f}{(1 - R_0^{-2})} \left(1 - \frac{2c_{33}}{c_{11} + c_{21}}\right) \tag{36}$$

$$P_f = \frac{1}{K_1} R_0^{K_1} \left[R_0^{-K_1} - 1\right] \tag{37}$$

9.4 NUMERICAL RESULTS AND DISCUSSIONS

The above investigations elaborate the initial yielding and fully plastic state of hollow cylinder subjected to external pressure. The cases of three cylinders were considered, first is a boron-aluminum fiber-reinforced orthotropic composite; second is the human femur bone and third is a cylinder made of transversely isotropic material, i.e., titanium (Figure 9.5).

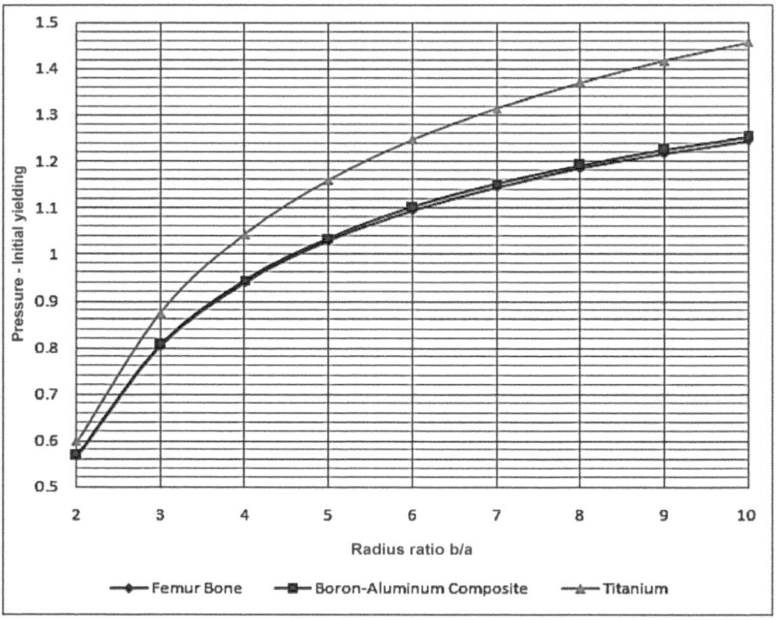

FIGURE 9.5 Pressure at initial yielding in the cylinder.

In Figure 9.5, curves are plotted for external pressure at initial yielding state and radii ratio R_0=b/a of the cylinders. The graph has been plotted for considerably thick cylinders. The curves show how the walls of the cylinders yield when various ratios of radial distances were considered. It is observed that the femur bone and the orthotropic composite yield in a very similar manner when the pressure is applied at the external surface, for the model with considerable thickness. The transversely isotropic material (titanium) yields at relatively lower magnitudes of pressure. Sufficient to say that the yielding of femur bone is highly similar to that of the composite material at various levels of pressure.

The radial, circumferential, and axial stresses which were calculated at initial yielding are plotted in Figures 9.6, 9.7, and 9.8 and fully plastic state are plotted in Figures 9.9, 9.10, and 9.11; along the radii ratio R = r/a.

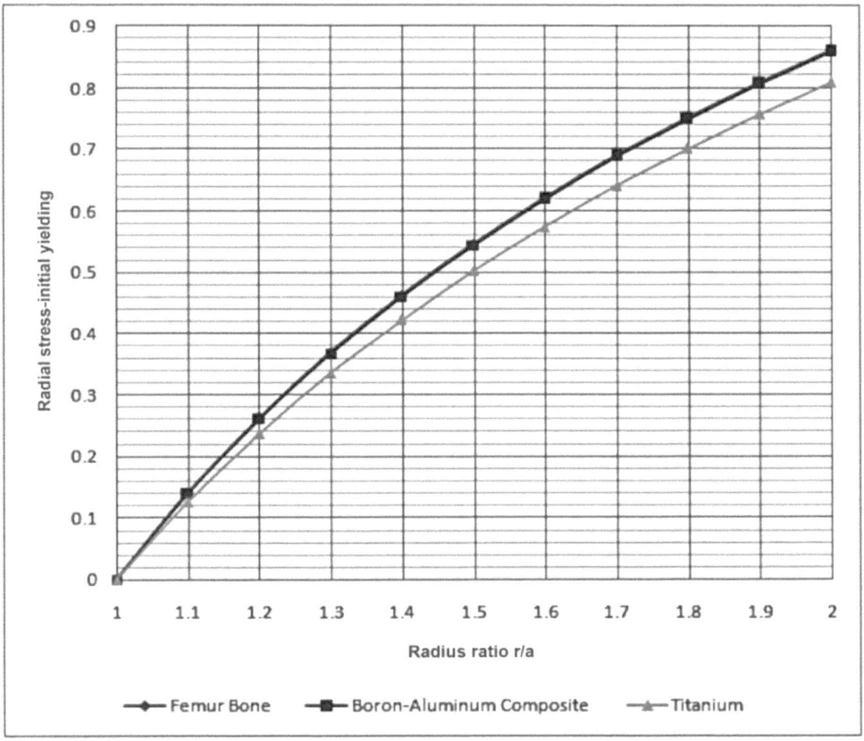

FIGURE 9.6 Radial stresses at initial yielding along the radius ratio r/a.

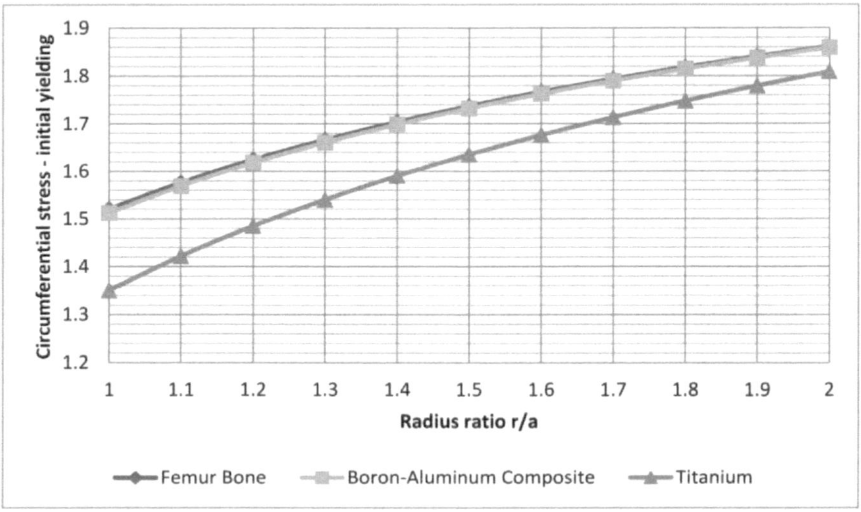

FIGURE 9.7 Circumferential stresses at initial yielding along the radius ratio r/a.

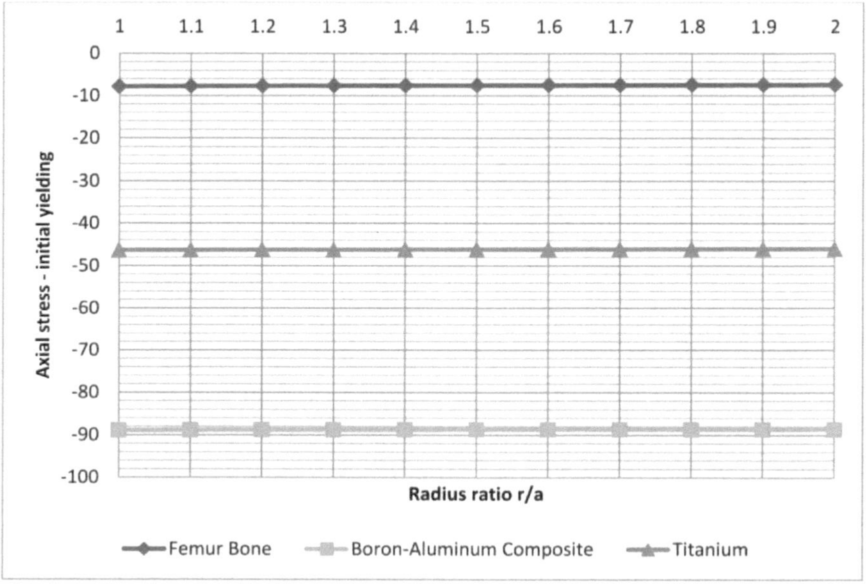

FIGURE 9.8 Axial stresses at initial yielding along the radius ratio r/a.

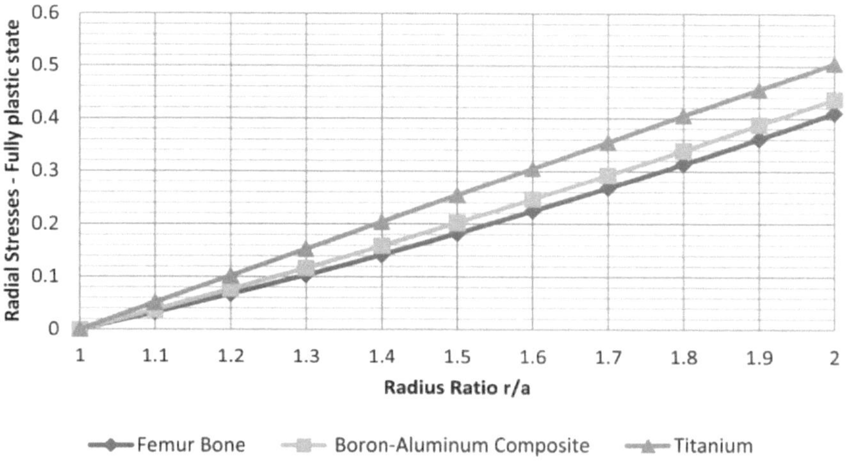

FIGURE 9.9 Radial stresses at fully plastic state along the radius ratio r/a.

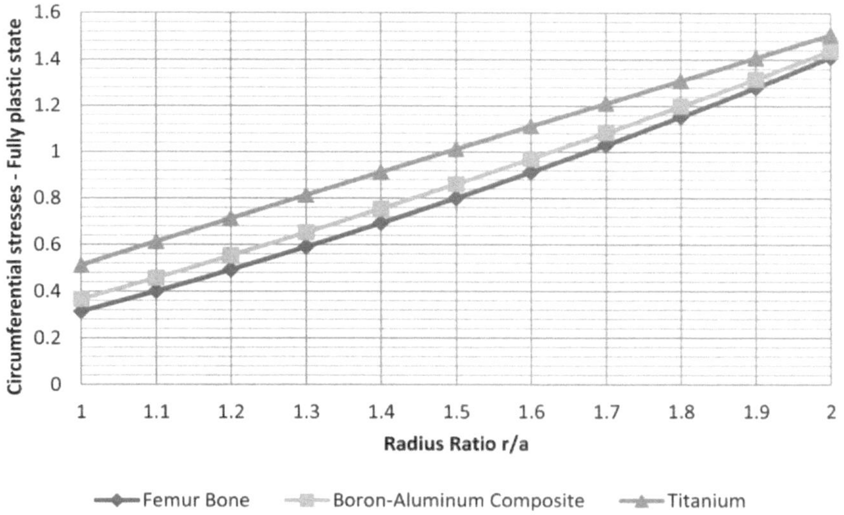

FIGURE 9.10 Circumferential stresses at fully plastic state along the radius ratio r/a.

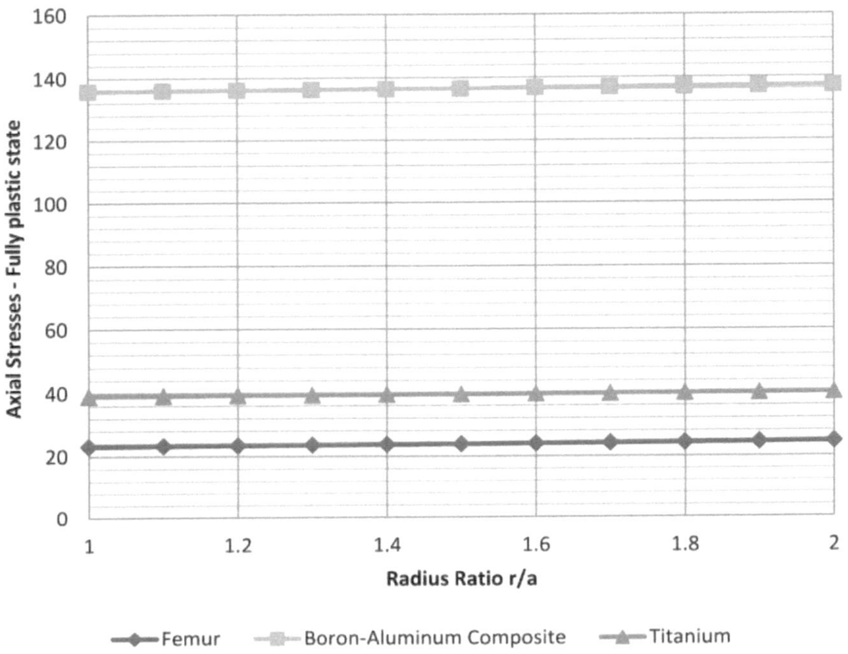

FIGURE 9.11 Axial stresses at fully plastic state along the radius ratio r/a.

It is observed that at the initial yielding stage, similar values of radial and circumferential stresses were observed for boron-aluminum composite and femur bone. Although titanium follows a similar trend for radial and circumferential stresses, yet the trends of boron-aluminum composite are much closer to that of the femur bone. For axial stresses, out of all three materials, boron-aluminum composite showed yielding at highest magnitudes of stress. The femur bone yielded at very low axial stresses followed by titanium. Similar trends were seen for the fully plastic state of all types of materials used.

9.5 CONCLUSIONS

Elastic-plastic stress concentrations have been determined in human femur bone, boron-aluminum composite and titanium subjected to external pressure using Seth's Transition Theory. It is sufficient to conclude that structures made of boron-aluminum orthotropic composite are likely to exhibit similar kind of deformation behavior when compared with human femur bone. Therefore, the use of this composite is secure from the design and manufacture perspective as compared to structures made of transversely isotropic material. The applications of boron-aluminum fiber-reinforced composite are diverse when we consider the design of prosthetic limbs and implants. The patients with bones damaged due to osteoporosis or any other traumatic incident such as sports injury or accident may be rehabilitated in a better way by implementing this piece of research.

ACKNOWLEDGMENTS

The authors gratefully acknowledge the experimental works of Prof. Dr. Hassel Ledbetter; Department of Mechanical Engineering, University of Colorado Boulder, Colorado, USA, and his team in the field of acoustic resonance spectroscopy.

KEYWORDS

- **bones**
- **boron-aluminum fiber reinforced composite**
- **cylinder**
- **femur**
- **implants**
- **orthotropic materials**
- **prosthetics**
- **titanium**

REFERENCES

1. Altenbach H., Altenbach J., & Kissing W., (2004). *Mechanics of Composite Structural Elements*. Springer-Verlag.
2. Bhatnagar, N. S., Kulkarni, P. S., & Arya, V. K., (1986). Steady state creep of orthotropic rotating discs of variable thickness. *Nucl. Eng. Des., 91*(2), 121–141.
3. Biswas, J. K., Rana, M., Majumder, S., Karmaka, R. S. K., & Roy, C. A., (2018). Effect of two-level pedicle-screw fixation with different rod materials on lumbar spine: A finite element study. *Journal of Orthopedics Science, 23*(2), 258–265.
4. Bundy K. J., (1974). Experimental studies of the non-uniformity and anisotropy of human compact bone. PhD Dissertation. Stanford University.

5. Ledbetter, H., Fortunko, C., & Heyliger, P., (1995). Orthotropic elastic constants of a boron-aluminum fiber-reinforced composite: An acoustic-resonance-spectroscopy study. *Journal of Applied Physics, 78*, 1542.
6. Ma, B. M., (1961). Creep analysis of rotating solid discs with variable thickness and temperature. *J. Frankl. Inst., 271*(1), 40–55.
7. Pitkin, M., Raykhtsaum, G., Pilling, J., Galibin, O. V., Protasov, M. V., Chihovskaya, J. V., et al., (2007). Porous composite prosthetic pylon for integration with skin and bone. *Journal of Rehabilitation Research and Development, 44*(5), 723–738.
8. Reilly, D. T., & Burstein, A. H., (1975). The elastic and ultimate properties of compact bone tissue. *Journal of Biomechanics, 8*, 393–405.
9. Roy, S., Khutia, N., Das, D., Das, M., Balla, V. K., Bandyopadhyay, A., & Chowdhury A. R., (2016). Understanding compressive deformation behavior of porous Ti using finite element analysis. *Materials Science and Engineering: C., 64*, 436–443.
10. Seth, B. R., (1962). Transition theory of elastic-plastic deformation, creep, and relaxation. *Nature, 195*, 896–897.
11. Seth, B. R., (1966). Measure concept in mechanics. *International Journal of Non-Linear Mechanics, 1*(1), 35–40.
12. Seth, B. R., (1972). Yield conditions in plasticity. *Arch. Mech. Strus., 24*(5), 769–776.
13. Shahi, S., Singh, S. B., & Thakur, P., (2019). Modeling creep parameter in rotating discs with rigid shaft exhibiting transversely isotropic and isotropic material behavior. *Journal of Emerging Technologies and Innovative Research, 6* (1), 387–395.
14. Singh, S. B., & Ray, S., (2001). Steady-state creep behavior in an isotropic functionally graded material rotating disc of Al-SiC composite. *Metall Trans., 32A*(7), 1679–1685.
15. Thakur, P., Shahi, S., Gupta, N., & Singh, S. B., (2017). Effect of mechanical load and thickness profile on creep in a rotating disc by using Seth's transition theory. *AIP Conf. Proc., 1859*(1), 20–24.
16. Tresca, H., (1868). M'emoire Sur I Ecoloment Descrops Solids.' *M'eoire presente's Par Divers Savents., 18*, 733–799.
17. Uyaner, M., Akdemir, A., Erim, S., & Avci, A., (2000). Plastic zones in a transversely isotropic solid cylinder containing a ring-shaped crack. *International Journal of Fracture, 106*, 161–175.
18. Gupta, V., & Singh S. B., (2016). Mathematical modeling of creep in a functionally graded rotating disc with varying thickness. *Regen. Eng. Transl. Med., 2*, 126–140.
19. Von, M. R., (1986). Mechanics of solids in the plastically deformable state. *NASA Tech. Memo., 1913*, 88488.
20. Van Buskirk, W. C., Stephen, C., & Ward, R. N., (1981). Ultrasonic measurement of orthotropic elastic constants of bovine femoral bone. *Journal of Biomechanical Engineering, 103*(2), 67–72.
21. Yoon, H. S., & Katz, J. L., (1976). Ultrasonic Wave propagation in human cortical bone—I. theoretical considerations for hexagonal symmetry. *Journal of Biomechanics, 9*, 407–412.

CHAPTER 10

CREEP MODELING IN A COMPOSITE ROTATING DISC WITH CONSTANT THICKNESS IN THE PRESENCE OF RESIDUAL STRESS

NISHI GUPTA[1] and SATYA BIR SINGH[2]

[1]*Department of Mathematics, UIS, Chandigarh University, Gharuan, India*

[2]*Department of Mathematics, Punjabi University, Patiala – 147004, India*

ABSTRACT

Residual stress significantly affects the engineering properties of materials and structural components notably fatigue life, distortion, dimensional, corrosion resistance, brittle fracture, and so forth. The thermal residual stresses induced due to thermal mismatch between the metal matrix and the ceramic reinforcement in metal matrix composite may impart plastic deformation to the matrix. In the present study, the steady-state creep has been investigated for composite rotating disc made of material 6061Al base alloy containing 20 vol% of SiC particle using isotropic/anisotropic Hoffman yield criterion and results are compared with those using von Mises yield criterion/Hill's criterion ignoring the difference in yield stresses. The creep behavior has been described by Sherby's constitutive model. Stress and strain rate distributions developed due to rotation have been calculated. It is concluded that the stress and strain distributions got affected from the thermal residual stress in an isotropic/anisotropic rotating disc, although the effect of residual stress on creep behavior in an anisotropic rotating disc is observed to be lower than those observed in an isotropic disc. Thus, the presence of residual stress in composite rotating disc with constant thickness needs attention for designing a disc.

10.1 INTRODUCTION

Residual stress expressively affects the engineering properties of materials and structural components, remarkably fatigue life, alteration, dimensional, oxidization, brittle fracture, etc. Residual stresses in a structural material are the structure of stresses which occur in a body in the absence of external loads. The presence of thermal residual stresses can induce the asymmetry in the tensile and compressive yield stresses of the composite. Residual stresses may be concentrated or eradicated by annealing, by plastic deformation or just by letting the piece at room temperature (RT) for adequate time. Because of its impact on the properties, the residual stress in composites has been the subject of several studies, both experimental and analytical [1, 17, 20]. Considering the importance of thermal residual stress, the objective of the present study has been to investigate the effects of the thermal residual stress on the stress distributions and the resulting creep deformation of the rotating disc made of silicon carbide particulate reinforced aluminum base composite. The investigation has been done for isotropic/anisotropic composite discs of constant thickness in presence/absence of thermal residual stresses by using isotropic Hoffman yield criterion. Results have been compared with those obtained using von Mises yield criterion/Hill's criterion ignoring difference in yield stresses. The creep behavior of the composite disc rotating at 15,000 rpm has been described by Sherby's constitutive model. The creep response of rotating disc is stated by a threshold stress with value of stress exponent as 8. It is determined that the presence of thermal residual stresses in composite rotating disc with constant thickness needs courtesy for designing a disc.

10.2 ISOTROPIC ANALYSIS

From symmetry considerations, principal stresses are in the radial, tangential, and axial directions for the Al-SiCp composite disc of uniform thickness, h, rotating with angular velocity, u. For the purpose of modeling, the following assumptions are made:

1. Steady-state condition of stress is assumed.
2. Elastic deformations are small and are neglected as compared to the creep deformations.
3. Biaxial state of stress exists at any point of the disc.
4. Composite shows steady-state creep behavior described by Sherby's constitutive model.

Singh and Ray [17] have used the following form of Hoffman yield criterion that employs uniaxial compressive and tensile yield stresses of f_c and f_t respectively,

$$f(\sigma_{ij})=(\sigma_{11}^2+\sigma_{22}^2+\sigma_{33}^2)-(\sigma_{11}\sigma_{22}+\sigma_{22}\sigma_{33}+\sigma_{33}\sigma_{11})+(f_c-f_t)(\sigma_{11}+\sigma_{22}+\sigma_{33})-f_c f_t=0 \quad (1)$$

The associated flow rule is

$$d\in_{ij}=\frac{\partial f}{\partial \sigma_{ij}}d\lambda \quad (2)$$

where $d\lambda$ is proportionality constant that must depend upon σ_{ij}, $d\sigma_{ij}$ and also on $\in_{ij}$. The effective stress and strain are related as follows:

$$d\in_e=\sigma_e d\lambda$$

The steady-state creep response of the Al-SiC$_p$ composite of varying composition is described in terms of Sherby's threshold stress based model given by,

$$\dot{\bar{\varepsilon}}=[M(r)(\bar{\sigma}-\sigma_0(r))]^8 \quad (3)$$

where $M=\frac{1}{E}\left[\frac{AD_L\lambda^3}{|\vec{b}_r|^5}\right]^{1/8}$ and $\bar{\sigma}$ are effective stress, $\dot{\bar{\varepsilon}}$ is effective strain rate, M is material creep constant, D_L is the lattice diffusivity, λ is the subgrain size, A is the constant, $|\vec{b}_r|$ is the magnitude of Burger's vector, E is the young's modulus and σ_0 is the threshold stress.

The values of creep parameters M and σ_0 have been obtained from the creep results reported for Al-SiC$_p$ composite Pandey et al. [16] and these values have been fitted by the following regression equations as a function of particle size (P), temperature (T) and volume content (V).

$$\ln M(r) = -34.91 + 0.2112 \ln P + 4.89 \ln T(r) - 0.591 \ln V \tag{4}$$

$$\sigma_0(r) = -0.02050\ P + 0.01378\ T(r) + 1.033\ V - 4.9695 \tag{5}$$

The variation of creep parameters in the rotating thermally graded discs along the radial distance has been determined in this study from the preceding equations for $P = 1.7\ \mu m$ and $V = 10 vol\%$

$$V = 10 vol\%.$$

$$\dot{\varepsilon}_r = \frac{\dot{\bar{\varepsilon}}}{2\bar{\sigma}} \{2\sigma_r - (\sigma_\theta + \sigma_z) + (f_c - f_t)\}$$

$$\dot{\varepsilon}_\theta = \frac{\dot{\bar{\varepsilon}}}{2\bar{\sigma}} \{2\sigma_\theta - (\sigma_z + \sigma_r) + (f_c - f_t)\}$$

$$\dot{\varepsilon}_z = \frac{\dot{\bar{\varepsilon}}}{2\bar{\sigma}} \{2\sigma_z - (\sigma_r + \sigma_\theta) + (f_c - f_t)\} \tag{6}$$

The effective stress using Hoffman yield criterion is given by,

$$\bar{\sigma} = \left\{ \frac{1}{\sqrt{2}} \left\{ (\sigma_\theta - \sigma_r)^2 + (\sigma_r - \sigma_z)^2 + (\sigma_z - \sigma_\theta)^2 + 2\,(f_c - f_t)(\sigma_z + \sigma_r + \sigma_\theta) \right\}^{1/2} \right\} \tag{7}$$

$\dot{\varepsilon}_r$, $\dot{\varepsilon}_\theta$, $\dot{\varepsilon}_z$ of and σ_r, σ_θ, σ_z are the strain rates and the stresses respectively in the direction r, θ and z. $\dot{\bar{\varepsilon}}$ is the effective strain rate.

For biaxial state of stress σ_r and σ_θ the equation becomes

$$\bar{\sigma} = \left\{ \frac{1}{\sqrt{2}} \left\{ (\sigma_\theta)^2 + (\sigma_r)^2 + (\sigma_r - \sigma_\theta)^2 + 2\,(f_c - f_t)(\sigma_r + \sigma_\theta) \right\}^{1/2} \right\} \tag{8}$$

Using Eqs. (13) and (8) in constitutive equations (6), we get

$$\dot{\varepsilon}_r = \frac{d\dot{u}_r}{dr} = \frac{[M(r)(\bar{\sigma} - \sigma_0(r))]^8 (2x(r) - 1)}{2\left[(x(r))^2 - x(r) + 1\right]^{1/2}} \tag{9}$$

$$\dot{\varepsilon}_\theta = \frac{\dot{u}_r}{r} = \frac{[M(r)(\bar{\sigma} - \sigma_0(r))]^8 (2 - x(r))}{2\left[(x(r))^2 - x(r) + 1\right]^{1/2}} \tag{10}$$

$$\dot{\varepsilon}_z = \frac{-[M(r)(\bar{\sigma} - \sigma_0(r))]^8 (x(r) + 1)}{2\left[(x(r))^2 - x(r) + 1\right]^{1/2}} \tag{11}$$

where $x(r) = \sigma_r(r) / \sigma_\theta(r)$ is the ratio of radial and tangential stress at any radius r. Eqs. (9) and (10) can be solved to obtain $\sigma_\theta(r)$ as given

$$\sigma_\theta(r) = \frac{(\dot{u}_a)^{1/8}}{M(r)} \psi_1(r) + \psi_2(r) \tag{12}$$

where

$$\dot{u}_a^{1/8} = \frac{\int_a^b M(r)\sigma_\theta dr - \int_a^b M(r)\psi_2(r)dr}{\int_a^b \psi_1(r)dr} \tag{13}$$

$$\psi_1(r) = \frac{\psi(r)}{\left[(x(r))^2 - x(r) + 1\right]^{1/2}}, \tag{14}$$

$$\psi_2(r) = \frac{\sigma_0(r)}{\left[(x(r))^2 - x(r) + 1\right]^{1/2}}, \tag{15}$$

$$\psi(r) = \left[\frac{2\left[(x(r))^2 - x(r) + 1\right]^{1/2}}{r(2 - x(r))} \exp \int_a^r \frac{\varphi(r)}{r} dr \right]^{\frac{1}{8}}, \tag{16}$$

and

$$\varphi(r) = \frac{(2x(r) - 1)}{(2 - x(r))}. \tag{17}$$

The equation of motion of a rotating disc of uniform thickness h, may be obtained by considering the equilibrium of an element in the disc confined between radial distances r and r + dr and an interval of angle between θ and $\theta + d\,\theta$. The equilibrium of the forces in the radial direction implies that

$$\frac{d}{dr}\left[r\sigma_r(r)\right]-\sigma_\theta(r)+\rho(r)\omega^2 r^2 = 0 \tag{18}$$

where r is the density of the composite. Knowing the tangential stress distribution $\sigma_\theta(r)$, values of $\sigma_r(r)$ can be obtained from above relation as follows:

$$\sigma_r(r) = \frac{1}{r}\int_a^r \sigma_\theta(r)\,dr - \rho\frac{\omega^2}{3r}(r^3 - a^3) \tag{19}$$

As the tangential stress, σ_θ the radial stress σ_r is determined by equations at any point within the composite disc. Then the strain rates $\dot{\varepsilon}_r$, $\dot{\varepsilon}_\theta$ and $\dot{\varepsilon}_z$ can be calculated from equations respectively. An iterative solution technique can be employed until the boundary conditions, $\sigma_r(a) =$ $= 0$ and $\sigma_r(b) = 0$ are satisfied.

The effect of thermal gradient on creep in an isotropic rotating disc in presence of residual stress has been studied. We have computed stress and strain rate distributions for a rotating disc with linearly varying temperature with and without residual stress. The results have been plotted as in Figures 10.1–10.8.

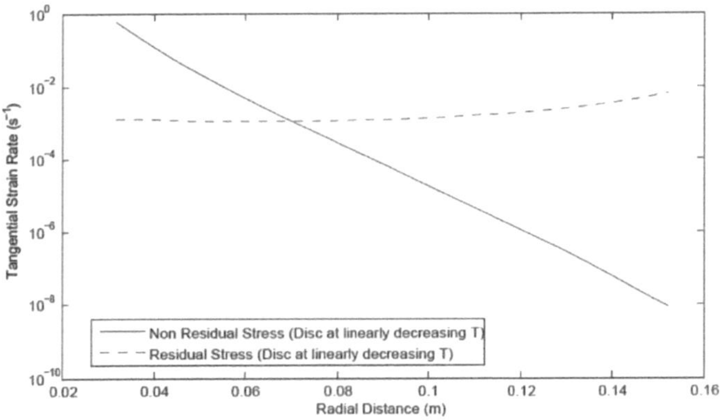

FIGURE 10.1 Variation of tangential strain rates along the radial distance in a disc at linearly decreasing temperature.

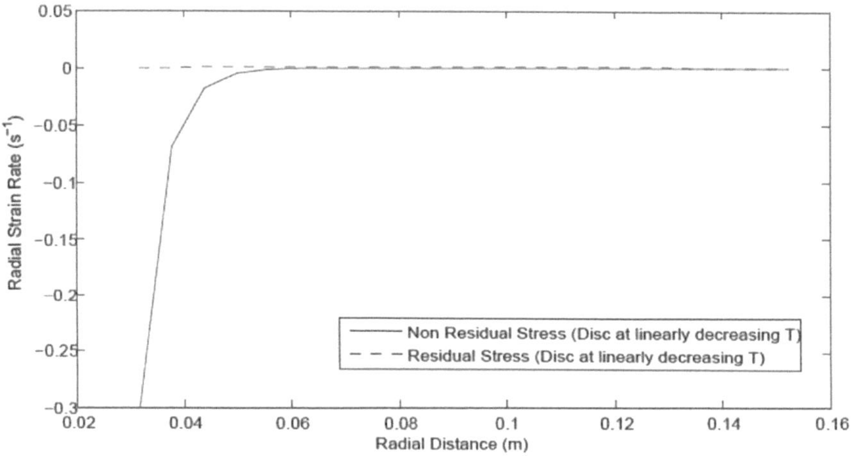

FIGURE 10.2 Variation of radial strain rates along the radial distance in a disc at linearly decreasing temperature.

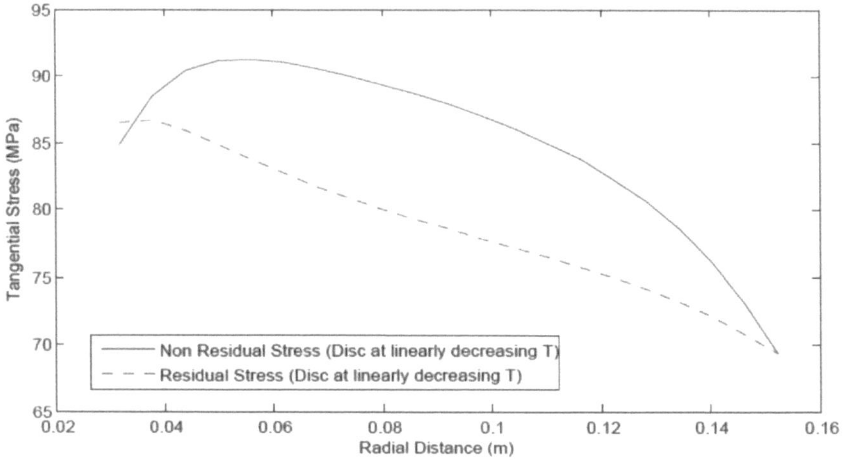

FIGURE 10.3 Variation of tangential stresses along the radial distance in a disc at linearly decreasing temperature.

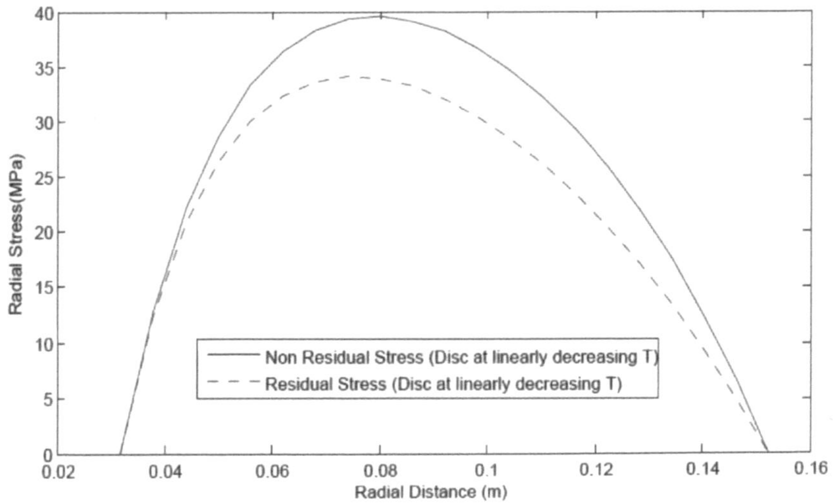

FIGURE 10.4 Variation of radial stresses along the radial distance in a disc at linearly decreasing temperature.

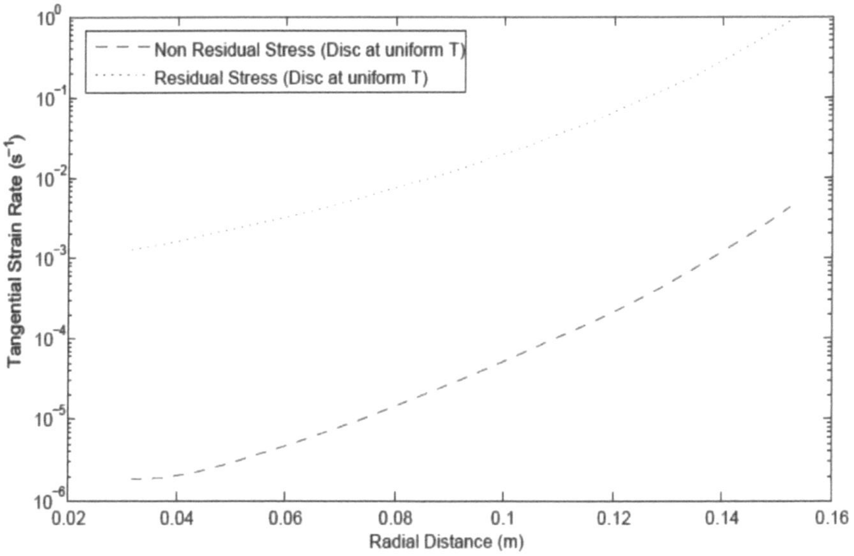

FIGURE 10.5 Variation of tangential strain rates along the radial distance in a disc at uniform temperature.

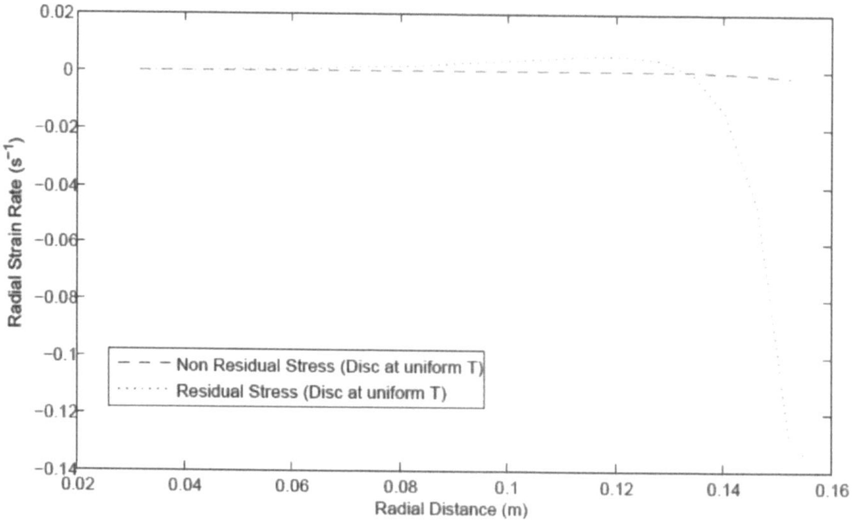

FIGURE 10.6 Variation of radial strain rates along the radial distance in a disc at uniform temperature.

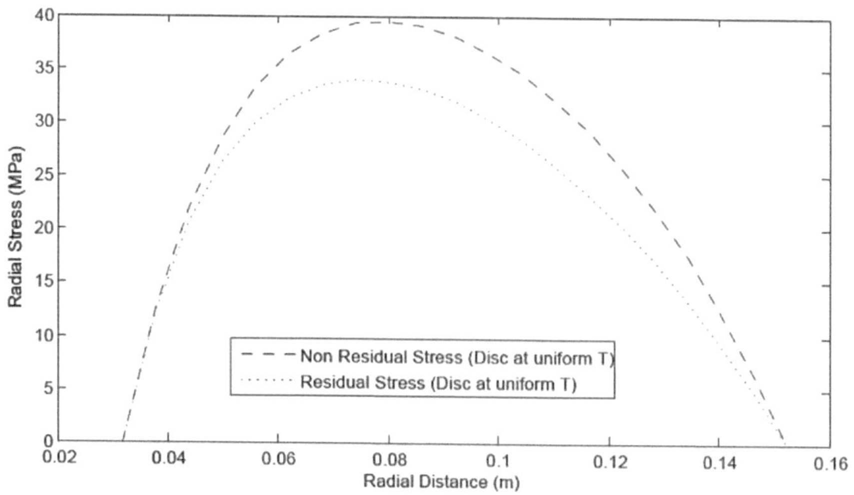

FIGURE 10.7 Variation of tangential stresses along the radial distance in a disc at uniform temperature.

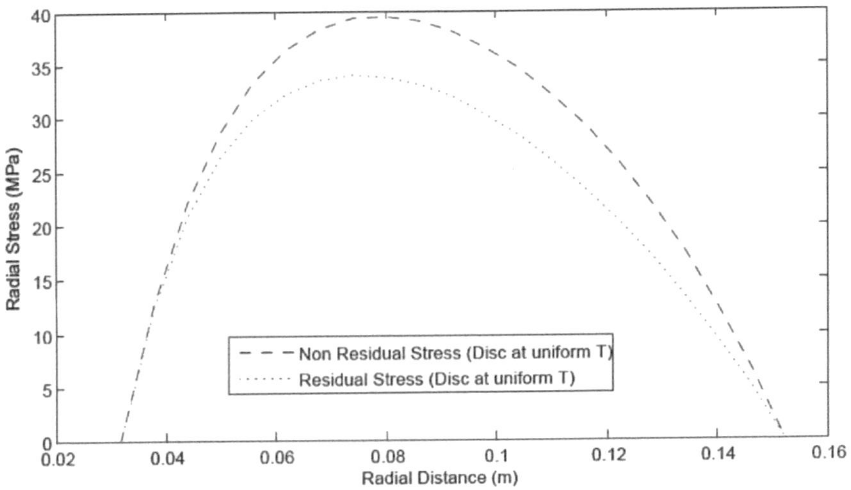

FIGURE 10.8 Variation of radial stresses along the radial distance in a disc at uniform temperature.

10.3 ANISOTROPIC ANALYSIS

For the purpose of creep analysis in disc with constant thickness, the following assumptions are made:

1.	Material of disc is orthotropic and incompressible.
2.	Elastic deformations are small for the disc and therefore they can be neglected as compare to creep deformation.
3.	Axial stress in the disc may be assumed to be zero as thickness of disc is assumed to be very small compared to its diameter.
4.	The composite shows a steady-state creep behavior.

$$\dot{\bar{\varepsilon}} = \left[M(r)(\bar{\sigma} - \sigma_0(r))\right]^8 \tag{20}$$

where $M = \frac{1}{E}\left[\frac{AD_L\lambda^3}{\left|\vec{b}_r\right|^5}\right]^{1/8}$ and $\bar{\sigma}$ are effective stress, $\dot{\bar{\varepsilon}}$ is effective strain rate, M is material creep constant, D_L is the lattice diffusivity, λ is the subgrain size, A is the constant, $\left|\vec{b}_r\right|$ is the magnitude of Burger's vector, E is the young's modulus and σ_0 is the threshold stress.

The different material combinations in the composite are conceptually replaced by an equivalent monolithic material that has the yielding and creep behavior similar to those displayed by the composite. Taking reference frame along the principal directions of r, θ and z, the generalized constitutive equations for an anisotropic disc under multiaxial stress condition are given as,

$$\dot{\varepsilon}_r = \frac{\dot{\bar{\varepsilon}}}{2\bar{\sigma}}\{(G+H)\sigma_r - H\sigma_\theta - G\sigma_z + (f_c - f_t)\} \tag{21}$$

$$\dot{\varepsilon}_\theta = \frac{\dot{\bar{\varepsilon}}}{2\bar{\sigma}}\{(H+F)\sigma_\theta - F\sigma_z - H\sigma_r + (f_c - f_t)\} \tag{22}$$

$$\dot{\varepsilon}_z = \frac{\dot{\bar{\varepsilon}}}{2\bar{\sigma}}\{(F+G)\sigma_z - G\sigma_r - F\sigma_\theta + (f_c - f_t)\} \tag{23}$$

where the effective stress, $\bar{\sigma}$, is given by

$$\bar{\sigma} = \left\{\frac{1}{G+H}\left[F(\sigma_\theta - \sigma_z)^2 + G(\sigma_z - \sigma_r)^2 + H(\sigma_r - \sigma_\theta)^2\right]\right\}^{1/2} \tag{24}$$

where F, G, H and are anisotropic constants the material, $\dot{\varepsilon}_r$, $\dot{\varepsilon}_\theta$, $\dot{\varepsilon}_z$ and σ_r, σ_θ, σ_z are the strain rates and the stresses respectively in the direction r, θ and z. $\dot{\bar{\varepsilon}}$ is the effective strain rate and $\bar{\sigma}$ is the effective stress and f_c, f_t are uniaxial compression and tensile yield stresses, respectively. For biaxial state of stress $(\sigma_r, \sigma_\theta)$, the effective stress is,

$$\bar{\sigma} = \left\{\frac{1}{(G+H)}\left\{F\,\sigma_\theta^2 + G\sigma_r^2 + H(\sigma_r - \sigma_\theta)^2\right\}\right\}^{1/2} \tag{25}$$

Using Eqs. (20) and (25), Eq. (21) can be rewritten as,

$$\dot{\varepsilon}_r = \frac{d\dot{u}_r}{dr} = \frac{\left[\left(\frac{G}{F}+\frac{H}{F}\right)x - \frac{H}{F} + \frac{f_c - f_t}{\sigma_\theta}\right]\left[M(r)\,(\bar{\sigma} - \sigma_0)\right]^8}{\sqrt{\frac{G}{F}+\frac{H}{F}}\left[\left(\frac{G}{F}+\frac{H}{F}\right)x^2 - 2\,\frac{H}{F}\,x + \left(\frac{G}{F}+\frac{H}{F}\right)\right]^{1/2}} \tag{26}$$

Similarly from Eq. (22) become

$$\dot{\varepsilon}_\theta = \frac{\dot{u}_r}{r} = \frac{\left[\left(1+\frac{H}{F}\right) - \frac{H}{F}x + \frac{f_c - f_t}{\sigma_\theta}\right]\left[M(r)\left(\bar{\sigma} - \sigma_0\right)\right]^8}{\sqrt{\frac{G}{F}+\frac{H}{F}}\left[\left(\frac{H}{F}+\frac{G}{F}\right)x^2 - 2\frac{H}{F}x + \left(1+\frac{H}{F}\right)\right]^{1/2}} \tag{27}$$

From the material's incompressibility assumption, it follows that

$$\dot{\varepsilon}_z = -\left(\dot{\varepsilon}_r + \dot{\varepsilon}_\theta\right) \tag{28}$$

where, $x = \frac{\sigma_r}{\sigma_\theta}$, is the ratio of radial and tangential stresses and $\dot{u}_r = du/dt$ is the radial deformation rate.

Dividing Eq. (26) by Eq. (27),

$$\varphi(r) = \frac{((G/F)+(H/F))\,x - (H/F) + ((f_c - f_t)/\sigma_\theta)}{(1+(H/F)) - (H/F)x + ((f_c - f_t)/\sigma_\theta)} \tag{29}$$

where,

$$\varphi(r) = \frac{d\dot{u}_r}{dr} \cdot \frac{r}{\dot{u}_r}$$

This implies,

$$\frac{d\dot{u}_r}{\dot{u}_r} = \frac{\varphi(r)}{r}dr$$

Integrating and taking limit a to r on both sides,

$$\dot{u}_r = \dot{u}_{r_i} \exp\int_a^r \frac{\varphi(r)}{r}dr \tag{30}$$

where, *r* is the radial deformation rate at the inner radius.

Dividing Eq. (30) by *r* and equated to Eq. (27),

$$\bar{\sigma} - \sigma_0(r) = \frac{(\dot{u}_{r_i})^{1/8}}{M(r)}\,\psi(r) \tag{31}$$

where,

$$\psi(\mathrm{r}) = \left\{ \frac{\sqrt{\frac{G}{F}+\frac{H}{F}}}{r} \cdot \frac{\left[\left(\frac{H}{F}+\frac{G}{F}\right)x^2 - \frac{2Hx}{F} + \left(1+\frac{H}{F}\right)\right]^{1/2}}{\left[\left(1+\frac{H}{F}\right) - \frac{H}{F}x + \frac{f_c - f_t}{\sigma_\theta}\right]} \exp.\int_a^r \frac{\varphi(\mathrm{r})\mathrm{dr}}{\mathrm{r}} \right\}^{1/8} \tag{32}$$

Substituting $\bar{\sigma}$ from Eq. (25) to Eq. (31), it gives,

$$\left\{\left(\frac{F}{G+F}\right)\left[\left(\frac{G}{F}+\frac{H}{F}\right)x^2 - 2\frac{H}{F}x + \left(\frac{H}{F}+1\right)\right]\right\}^{1/2} \sigma_{\grave{e}}(r) - \sigma_0(r) = \frac{\left(\dot{\mathrm{u}}_{r_i}\right)^{1/8}}{\mathrm{M(r)}}\psi(\mathrm{r}) \tag{33}$$

This implies,

$$\sigma_\theta(r) = \frac{\left(\dot{u}_{r_i}\right)^{1/8}}{M(r)}\,\psi_1(r) + \psi_2(r) \tag{34}$$

where,

$$\psi_1(r) = \frac{\psi(r)}{\left\{\left(\frac{F}{G+H}\right)\left[\left(\frac{G}{F}+\frac{H}{F}\right)x^2 - 2\frac{H}{F}x + \left(1+\frac{H}{F}\right)\right]\right\}^{1/2}} \tag{35}$$

and

$$\psi_2(r) = \frac{\sigma_0}{\left\{\left(\frac{F}{G+H}\right)\left[\left(\frac{G}{F}+\frac{H}{F}\right)x^2 - 2\frac{H}{F}x + \left(1+\frac{H}{F}\right)\right]\right\}^{1/2}} \tag{36}$$

The equation of equilibrium for a rotating disc with constant thickness can be written as:

$$\frac{d}{dr}\left[r\sigma_r(r)\right] - \sigma_\theta(r) + \rho\omega^2 r^2 = 0 \tag{37}$$

Integrating Eq. (36) from a to b and putting the boundary conditions $\sigma_r = 0$ at $r = a$ and $\sigma_r = 0$ at $r = b$, we get

$$\int_a^b \sigma_\theta dr = \rho\omega^2(b^3 - a^3)/3 \tag{38}$$

In the first iteration, $\sigma_\theta = \sigma_{\theta_{avg}}$, where $\sigma_\theta = \sigma_{\theta_{avg}}$ is the average tangential stress over the cross-section of the disc. Therefore, Eq. (33) in the first iteration may be written as:

$$\dot{u}_{r_i}^{1/8} = \frac{\sigma_{\theta_{avg}}\int_a^b M(r)dr - \int_a^b M(r)\psi_2(r)dr}{\int_a^b \psi_1(r)dr} \tag{39}$$

The values of $\sigma_r(r)$ can be as obtained by integrating Eq. (37) from a to$r$ as follows:

$$\sigma_r(r) = \frac{1}{r}\int_a^r \sigma_\theta(r)dr - \frac{\rho\omega^2(r^3 - a^3)}{3r} \tag{40}$$

Knowing the values of $\sigma_\theta(r)$ from Eq. (34), the radial stress, σ_r is determined by Eq. (40) at any point within the composite disc and the strain rates $\dot{\varepsilon}_r$ and $\dot{\varepsilon}_\theta$ are calculated from Eqs. (26) and (27), respectively.

10.4 VALIDATION

In order to validate the analysis and the developed computer program, the results for a rotating steel disc were obtained and compared with the available experimental results of Wahl et al. [21] for the same type of disc and operating under same conditions. The comparison of present theoretical study and experimental results of Wahl et al. [21] for strain rates versus radial distance is shown graphically in the figures below. This graph depicts that there is a good agreement between the present theoretical and Wahl's experimental results for radial strain rates versus radial distance. The tangential strain rates also follow the same trend for present as well as experimental results of Wahl throughout the radial distance as one move from inner to outer radii. A good agreement and similar trends observed between the present theoretical and the Wahl's experimental strain rates inspires the confidence in the computer program developed (Figures 10.9–10.14).

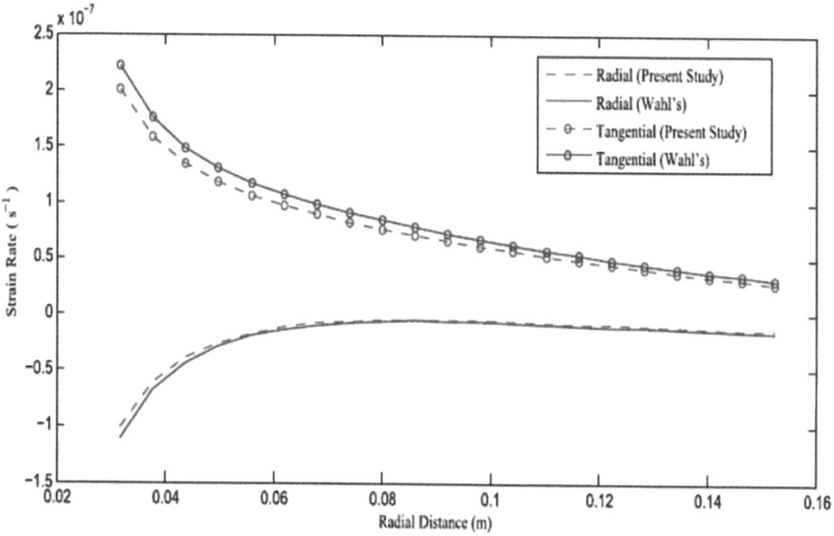

FIGURE 10.9 Theoretical results vs. experimental results of Wahl et al. [21].

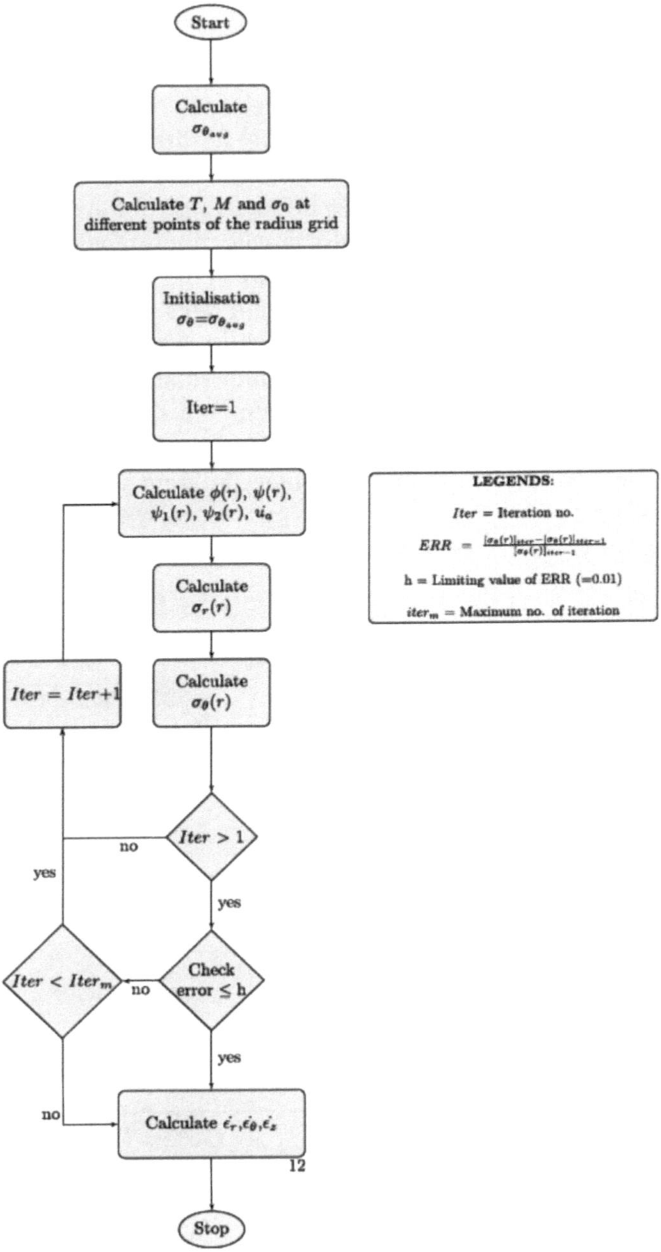

FIGURE 10.10 Numerical scheme of computation.

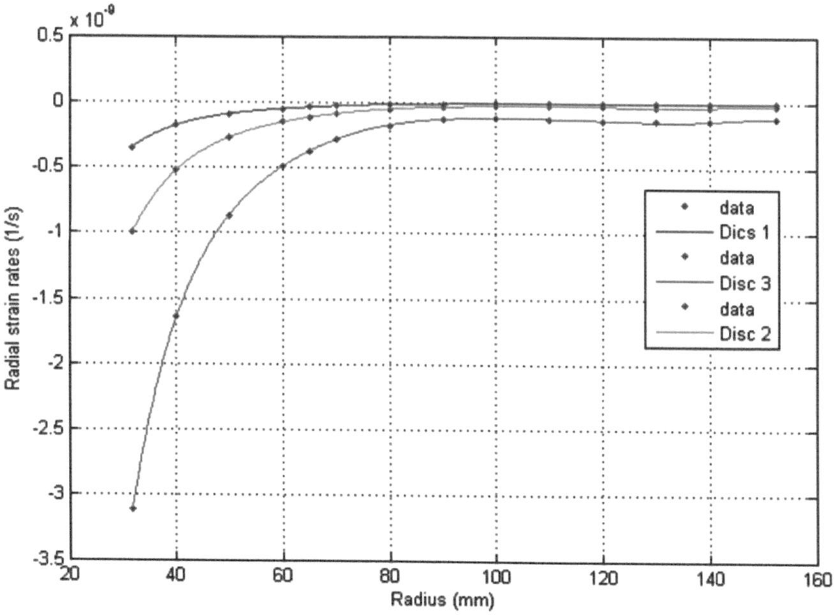

FIGURE 10.11 Variation of radial strain rates along the radial distance in the isotropic/anisotropic discs rotating with an angular velocity 15000 rpm at 616K.

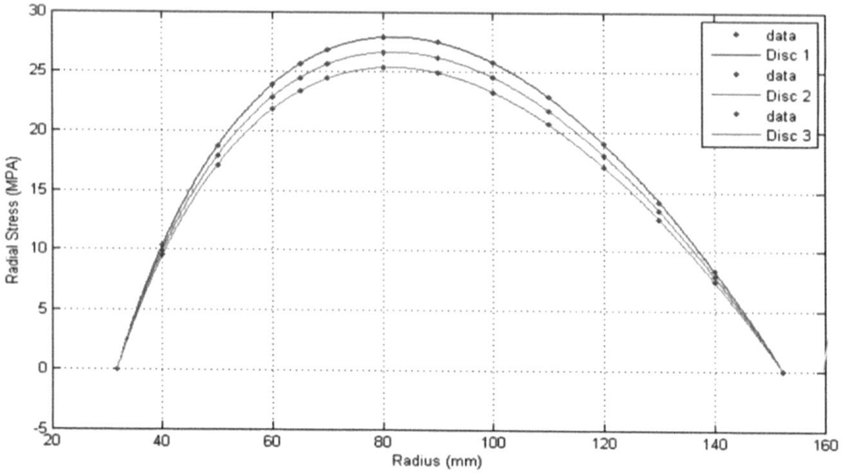

FIGURE 10.12 Variation of radial stress along the radial distance in isotropic/anisotropic discs rotating with an angular velocity 15000 rpm at 616K.

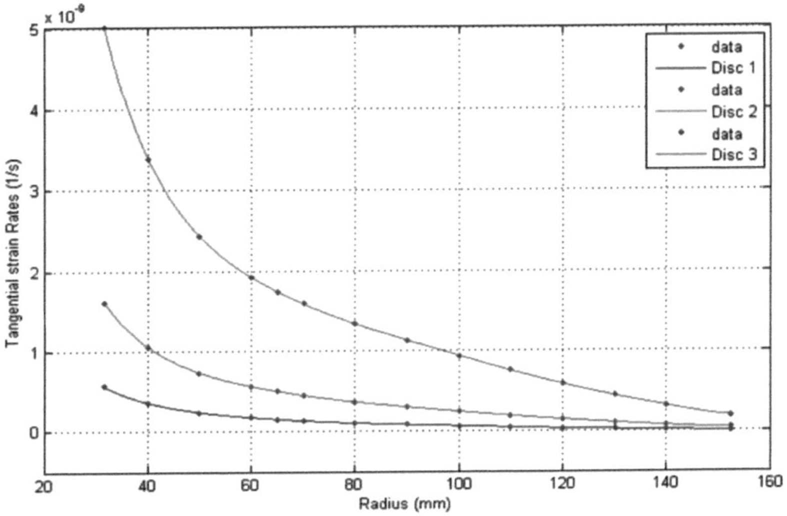

FIGURE 10.13 Variation of tangential strain rate along the radial distance in isotropic/anisotropic discs rotating with an angular velocity 15000 rpm at 616K.

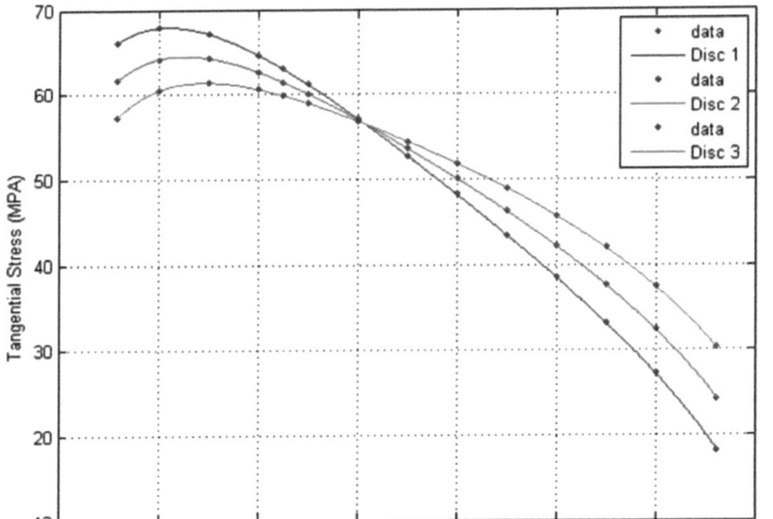

FIGURE 10.14 Variation of tangential stress along the radial distance in isotropic/anisotropic discs rotating with an angular velocity 15000 rpm at 616.

Figure 10.12 shows the variation of radial stresses along the radius of the isotropic/anisotropic rotating disc in presence/absence of residual stress. It is observed that in presence of residual stress, the radial stress developing due to rotation is slightly lesser than the radial stress of an isotropic/anisotropic disc without residual stress, although the change in the magnitude of radial stress distribution is very small in the isotropic discs with constant thickness due to presence of residual stress. As one moves from the inner radius to the outer radius of the disc, the radial stress increases from zero and reaches a maximum near the middle region of the disc with constant thickness and then starts decreasing towards the outer region. Figure 10.13 shows that in presence of tensile residual stress, the tangential strain rates enhance significantly in both the isotropic discs compared to the discs without residual stress. Also, the difference in the tangential strain rate caused due to presence and absence of residual stresses (i.e., the residual effect) goes on increasing with radial distance and the extent of increase in difference is maximum in the region near the outer radius in both the isotropic/anisotropic discs having constant thickness. Secondly, it is also noticed that variation in magnitude due to residual stress in an anisotropic disc is smaller compared to that for an isotropic disc. In an isotropic/anisotropic disc with/without residual stress, the tangential strain rates are highest at the inner radius and then decrease continuously, when one moves towards the outer radius of the disc. The trend of variation of tensile strain rate in tangential direction remains the same in an isotropic/anisotropic disc in the presence/absence of residual stress, but the magnitude can be reduced in an anisotropic disc. Figure 10.14 shows that in the presence of residual stress, the radial strain rate in an isotropic/anisotropic rotating disc with linear thickness and results are compared with those for without residual stress. By seeing both the discs of isotropic/anisotropic material, it is concluded that in the absence of residual stress, the nature of the radial strain rate which was compressive, becomes tensile at the middle of the disc in the presence of residual stress. Another point to be observed is that the magnitude of radial strain rate firstly increases rapidly with radial distance and then starts decreasing. It reaches a minimum before increasing again towards the outer radius in both the isotropic/anisotropic discs with residual/without residual stress.

10.5 RESULTS AND DISCUSSION

A computer program based on the analysis has been developed to obtain the stress distributions and strain rate of the rotating constant thickness discs containing silicon carbide particle in a matrix of pure aluminum in presence of thermal residual stress and obtained results are compared to the disc without residual stress to analyze the importance of residual stress. The parameters used in this paper for the analysis has been given in Table 10.1. Once again, a good agreement is noticed between the results obtained from the current investigation and those obtained experimentally by Wahl et al. [21]. For the analysis, the tensile residual stress ($\Delta\ \sigma_y$) has been taken as 32 MPa [18, 19]. The creep analysis in a rotating disc made of Al-SiC particle composite having constant thickness shown in Figure 10.1, has been carried using isotropic Hoffman yield criterion of yielding and results are compared with those using von Mises yield criterion/Hill's criterion of yielding ignoring difference in yield stresses i.e., ($\Delta\ \sigma_y = 0$) Figure 10.11 shows the tangential stress in an isotropic/anisotropic rotating disc with constant thickness in presence of residual stress represented by Disc 1, Disc 2 and Disc 3 and the results are compared with those for without residual stress. It is concluded that in the isotropic, the tangential stress is a little lower in region near the inner radius and slightly higher in region near the outer radius in the presence of residual stress as compared to the disc without residual stress. It is also noted that the effect of residual stress on tangential stress distribution is lesser in an anisotropic rotating disc of constant thickness compared to an isotropic disc of linearly constant thickness. Also, the tangential stress becomes more uniform in the presence/absence of residual stress in an anisotropic disc compared to an isotropic disc.

TABLE 10.1 Parameters and Operating Conditions for Steel Disc

Parameters for steel disc:
Density of disc material $\rho = 2812.4\, kg/m^3$
Inner radius of disc, $a = 31.75\ mm$
Outer radius of disc, $b = 152.4\ mm$
Particle size, $P = 1.7\ \mu m$
Particle content, $V = 20\%$
Creep parameters: $M = 5.83\times10^{-3}\ s^{-1/8}\ /\ MPa$ and $\sigma_0 = 24.44\ MPa$

Young's modulus, $Al = 70\ GPa$
Young's modulus, $SiC = 47\ GPa$
Density, $Al = 2713\ kg/m^3$
Density, $SiC = 3210\ kg/m^3$
The stress exponent of disc $n = 8$
The ratio of isotropic constants, $G/F = 1$, $H/F = 1$
The ratio of anisotropic constants, $G/F = 1.34$, $H/F = 1.64$
Operating conditions:
Angular velocity of Disc, $\omega = 15{,}000\ rpm$
Operating temperature, $T = 616\ K$
Creep duration, $t = 180\ hrs$

10.6 CONCLUSION

It can be concluded that residual stress may cause significant distortion in an isotropic/anisotropic rotating disc having constant thickness, but the magnitude of distortion can be reduced by creating anisotropy during processing.

KEYWORDS

- **anisotropy**
- **metal matrix composite**
- **residual stress**
- **rotating disc**
- **safe design**
- **tensile residual stress**

REFERENCES

1. Arsenault, R. J., & Taya, M., (1987). Thermal residual stress in metal matrix composite. *Acta Metallurgica, 35*(3), 651–659.

2. Audebert, N., Barber, J. R., & Zagrodzki, P., (1998). Bucking of automatic transmission clutch plates due to thermo elastic-plastic residual stresses. *Journal of Thermal Stresses, 21*, 309–326.
3. Autar, K. K., (1997). *Mechanics of Composites Materials* (pp. 2–18). CRC Press Boca Raton, New York.
4. Bache, M. R., Evans, W. J., & Uygur, I., (1998). Fatigue life prediction for notch geometries in particle reinforced metal matrix composites. *Material of Science and Technology, 14*, 1065–1069.
5. Bektas, N. B., (2005). Elastic-plastic and residual stress analysis of a thermoplastic composite hollow disc under internal pressures. *Journal of Thermoplastic Composite Materials, 18*, 363–375.
6. Bhatnagar, N. S., Kulkarni, M. P. S., & Arya, V. K., (1986). Steady-state creep of orthotropic rotating disks of variable thickness. *Nuclear Engineering and Design, 91*(2), 121–141.
7. Chamoli, N., Rattan, M., & Singh, S. B., (2010). Effect of anisotropy on the creep of a rotating disc of al-SiCp composite. *International Journal of Contemporary Mathematical Sciences, 5*(11), 509–516.
8. Davis, L. C., & Allison, J. E., (1993). Residual stresses and their effects on deformation." *Metallurgical Transactions A, 24*(11), 2487–2496.
9. Gupta, V. K., Singh, S. B., Chandrawat, H. N., & Ray, S., (2004). Steady state creep and material parameters in a rotating Disc of Al-SiC_p Composite. *European Journal of Mechanics, A/Solids, 23*(2), 335–344.
10. Gupta, V., & Singh, S. B., (2011). Modeling anisotropy and steady state creep in a rotating disc of Al-SiC_p having varying thickness. *International Journal of Scientific and Engineering Research, 2*(10), 1–12.
11. Gupta, V., & Singh, S. B., (2012). Creep modeling in a composite rotating disc with thickness variation in presence of residual stress, Hindawi publishing corporation. *International Journal of Mathematics and Mathematical Sciences*, Article ID 924921.
12. Hosford, & William, F., (2005). *Residual Stresses in Mechanical Behavior of Materials* (pp. 308–321). Cambridge University Press, ISBN 978-0-521-84670-7.
13. Hu, G. K., & Huang, G. L., (2000). Influence of residual stress on the elastic-plastic deformation of composites with two- or three-dimensional randomly oriented inclusions. *Acta Mechanica, 141*(3), 193–200.
14. Jahed, H., & Shirazi, R., (2001). Loading and unloading behavior of a thermoplastic disc. *International Journal of Pressure Vessels and Piping, 78*(9), 637–645.
15. Mishra, R. S., & Pandey, A. B., (1990). Some observations on the high-temperature creep behavior of (6061). Al-SiC composites. *Metallurgical Transactions A, 21*(7), 2089–2090.
16. Pandey, A. B., Mishra, R. S., & Mahajan, Y. R., (1992). Steady state creep behavior of silicon carbide particulate reinforced aluminum composites. *Acta Metallurgica Et Materialia, 40*(8), 2045–2052.
17. Singh, S. B., & Ray, S., (2003). Newly proposed yield criterion for residual stress and steady state creep in an anisotropic composite rotating disc. *Journal of Materials Processing Technology, 143–144*(1), 623–628.
18. Singh, S. B., & Ray, S., (2004). Modeling the creep in an isotropic rotating Disc of Al-SiC_w composite in presence of thermal residual stress. *Proceedings of the 3rd*

International Conference on Advanced Manufacturing Technology (ICAMT '04), (pp. 766–770). Kuala Lumpur, Malaysia.

19. Singh, S. B., (2008). One parameter model for creep in a whisker reinforced anisotropic rotating disc of Al-SiC_w composite. *European Journal of Mechanics, A/ Solids, 27*(4), 680–690.
20. Singh, S. B., & Rattan, M., (2010). Creep analysis of an isotropic rotating Al-SiCp composite disc taking into account the phase-specific thermal residual stress. *Journal of Thermoplastic Composite Materials, 23*(3), 299–312.
21. Wahl, A.M. (1954). Stress Distribution in Rotating Discs Subjected to Creep at Elevated Temperature, *Journal of Applied Mechanics, ASME Transactions, 29*, 299–305.
22. Xuan, F. Z., Chen, J. J., Wang, Z., & Tu, S. T., (2009). Time-dependent deformation and fracture of multi-material systems at high temperature." *International Journal of Pressure Vessels and Piping, 86*(9), 604–615.

CHAPTER 11

BIG DATA DEVELOPMENT PLATFORM FOR ENGINEERING APPLICATIONS

HERU SUSANTO[1] and FANG-YIE LEU[2]

[1]*School of Business, University Technology of Brunei, Brunei Darussalam Information Management, Tunghai University, Taiwan Research Center for Informatics, The Indonesian Institute of Sciences, Indonesia, E-mail: heru.susanto@lipi.go.id*

[2]*Computer Science, Tunghai University, Taiwan, E-mail: leufy@thu.edu.tw*

ABSTRACT

In today's modern society, information systems (IS) and big data can be found everywhere. It is so common that higher learning institutions have also implemented them in the academic field. Many improvements in the education system have been made by using IS and big data. They help increase the productivity of stakeholders in the academic world and also, their level of efficiency. This chapter will discuss how IS and big data are applied in engineering. It will focus on how big data transformed the engineering system, its importance, the application of it in the academic field and the challenges it faces as a whole.

11.1 INTRODUCTION

An information system (IS) is a group of computer tools, which can be used for collecting, storing, or processing data. Institutions depend on these systems to help enable and support their operations, interact with people, manage their workforce, provide services and many other processes. Every day, the amount of data collected in the digital tools grows tremendously than the previous years. As the amount of data increases, the use of IS becomes more essential and thus, with every passing year, the price of IS hardware has decreased significantly as well. This development is occurring under Moore's law as the heart of microprocessors has been doubling every 18–24 months [47]. Furthermore, the services of computer storage have been shifting from hardware to cloud systems. This is because of the rising concern towards the environmental impacts of the use of electric power by computer hardware.

Many ISs are mainly platforms to deliver data stored in databases. A database is a collection of interrelated data organized [47]. Databases store organized data so people can retrieve them from different criteria easily. Databases help support everyday operations and management of academic institutions.

To manage these data, we depend on IS to create, control, store, distribute, locate, and access this information. However, traditional computing solutions are not scalable enough to manage such a magnitude of data. These large sets of data volumes are known as big data. Big data is the result of collecting a large volume of data across many sites. It is important to have the right IS to cater high volume of data collected in a system. Academic institutions can benefit from big data if they know how to manage it accurately.

Big data is often determined according to the "3Vs" which is volume, variety, and velocity [36]. The 3Vs can also be described as the quantity, diverse types of data and rate of flow of information going into organizations that exceed the capacity of a traditional computer. The volume aspect of 3V is the size of data. It is difficult to determine the limits of big data, so this aspect is very relative in the education field. Variety is the different types or formats of big data. Therefore, this means that big data in academic collects, analyzes, and provides information with different backgrounds to ensure better learning resources for institutions. Lastly, Velocity is the increasing flow of data and the need for hardware in ISs to carry more and more information, and for software to process these data as quickly as possible. Big data ensures that stakeholders in the academic environment can have a quick access to information needed in their educational processes.

Today, there are many big data technologies created to help tackle these issues. They transform how big data can be analyzed and utilized correctly. The concept of big data is also expected to change the way of learning approaches, for example, e-learning - where more interactions of students and lecturers are encouraged. ISs and big data are intertwined as to make the academic system a more progressive institution. This will be discussed further below.

11.2 LITERATURE REVIEW

11.2.1 THE USE OF TECHNOLOGY

Information strategy is the support system for both academic strategy and IS strategy. It is helpful in terms of identifying types of information required, where to apply the information in order to aid and facilitate the main activities, or important goals of the strategy for the academic institution [24]. It is also needed to identify how valid and relevant the critical hypothesis are behind every academic strategy within the context of changing the environment and perceptions.

Many developments have been made in the academic institutions. This is in terms of implementing ISs in the academic settings. According to Shafique and Mahmood [45], there are many participants involved in an academic process where each user has their own needs to acquire certain information. These participants are amongst teaching staff, students, administration officers, researchers, and the general public. To accommodate these large numbers of stakeholders, several ISs are introduced, such as the UNESCO National Education Statistical Information Systems (NESIS) development Program and the education management information system (EMIS) to help countries to "systematically organize information related to the management of educational development" [45].

These systems help to solve common problems like precision, completeness, and accessibility of information that are usually faced by stakeholders. Other than that, ISs can also help in enhancing decision-making that is associated with the performance and development of students in class. Most importantly, these students' related information can be stored and be easily accessed in the system that can be of use again in the long run.

Some examples of ISs that are education-based are namely, the student management system, teacher management IS and the school management system. The Student Management System keeps details of students, such as their demographic data, class status and many more. Each student is linked to the institution with a unique code. Using this information, the system can track the student in any given institution via the unique code. Whereas, teacher management IS is used for tracking the induction, lecturer's training and professional development progress. This is to solve the problems with lecturers that have poor qualifications. Lastly, the School Management System keeps details of school's location, types of school, numbers of classrooms, bathrooms, library, and others [32].

Using computer programs such as Microsoft Data or Excel also improves the efficiency of data management and data manipulation. It is good to implement technology in academic because of the use of multimedia, flexibility, and real-time engagement. Multimedia, in this case, demonstrates a more enriched teaching and learning enterprise [29]. Technology also adds flexibility in teachers on how they present new knowledge and to get feedbacks from their students. For example, when students are asked to demonstrate their learning through multimedia presentations in class, their skills and content of knowledge will be enhanced. There can also be real-time engagement as students can explore from outside their homeland. For example, they could have video conferencing with other students in different countries to study and share about their new knowledge.

Although there is good in using technology in the academic context, there are also reasons to not implement it. An example would be like how not everyone is fond of using ISs in managing academic. According to Marcella and Knox [37], from their research survey, it was found that over 60% of university staff from the universities under their study reported that there were problems in the current computer interfaces and/or systems. The problems that were identified included the lack of expertise in the data manipulation or use of applications and the lack of knowledge on how or where data is 'housed' in terms of data sets inter-relations.

On the departmental level, it is acceptable to receive grants to develop new learning technologies. On the university scale as a whole organization, on the other hand, it would require for the state or government to provide a certain budget to implement new technology. Massive scale of technology implementation would require in building state-of-the-art classrooms to allow the use of the latest technology. Students who do not have access to personal computers will also experience problems in the using the latest technology, as they would have to use computers in campus to do their assignments and other course-related materials. This will hinder the progress of the students' performance.

11.2.2 IS IN HIGHER LEARNING INSTITUTIONS

Most important volume of information is mainly available in social media sites and media networks, however, the percentage of useful information is reduced as compared to other data that are readily available and provided by education institutions and business organizations. Big data in the academic perspective throughout the whole learning process is usually collected by a variety of management ISs. They are very effective and responsive in enabling academic experts to create and deliver knowledge, manage content of the website software, monitor participation and also, in assessing performance among learners. A Learning Management System, for example, is a web-based software application used in the academic environment for delivering knowledge online, which can be considered as part of e-Learning.

The term E-learning, i.e., online learning, according to Harasim [48] is publicly acknowledged to have a background of 'access' from the years in 1980s, but it does not have its beginnings exposed. Particularly, Kerry et al. [49] do not agree with the other authors who interpret e-Learning firmly as technological tools as being accessible that are web-base, web-distributed or web-capable. This is excellent, as generations today want a more interactive learning. Unlike traditional learning, students from this generation want interactivity.

E-learning, in general, is ranged inclusive from how emails are used between students and lecturers to having a whole learning class online or web-based. It is a solution that allows accessibility to training that became essential in complementing with the traditional way of teaching (face-to-face teaching). In complementing with traditional learning experiences, lecturers can still teach their lessons physically in the classroom by also incorporating the use of technology from time to time, such as activities that are done online using the Internet, simulations, virtual laboratories, and online testing [25].

Learning management systems (LMS) is usually used for conducting online courses and other aspects of learning in academic, but it also changes the term e-Learning in identifying how mechanisms used in the learning experiences to be delivered [38]. It can also be called as course management systems (CMS) and virtual learning environments (VLE). Ebardo and Valderama [50] described LMS as a software used to deliver, track, and also manage learning instructions. It also helps in many institutions in carrying out courses over the Internet and in featuring online collaborations. However, according to them, an LMS is not intended to train or develop, but to manage. This is because the architecture built for the system is required for managing and administrating training is not the same with the framework that is needed to instruct and also learn.

These LMSs come with both advantages and disadvantages. When the right learning strategies are implemented, LMS can increase the motivation of students, learning, class participation, feedbacks, and support during the process of learning. Another advantage is that the system supports all kinds of formats, like multimedia, video, and text. Materials of courses can be accessed at any time as they are available online. Materials on the courses are updated and students can see the changes made in the course materials. Educators can also modify information in accordance to the students' needs.

However, LMS can be more of a course-centered rather than student-centered. This is because LMS does not support various teaching styles. There is no guarantee that learning can be improved. Also, some educators have weak computer and information literacy skills and these skills are needed to use LMS successfully in supporting their teaching. According to Samsonov and Beard [51] teaching staff must learn how to operate within the environments and develop critical perspective of their use of the LMS in teaching (as cited in Ref. [44]). Additionally, many teachers also find it difficult in designing and organizing a mixture of learning activities that are appropriate with the students' needs, teaching skills and teaching styles [52].

An example of a learning management system used in higher learning institutions is Moodle. The Moodle is a platform used in learning that is based on the function and work of a teacher's principles that provide a capacity in education for educators to be able to design their content of teaching flexibly online and also, to be able to collaborate projects to experience constant feedback from students. Running Moodle requires no costs, therefore, higher learning institutes can adopt and support Moodle by themselves entirely if they aim to save budget. Further examples of LMS are BlackBoard, WebCT, and ToolBook.

With the help of ISs, records of revision histories and operational data are also stored efficiently. These data storage can help in decision support and to prepare institutions from making the wrong decision on certain subjects. Additionally, by having accurate and up-to-date information available, institutions can give a better quality service in education. Relationships with consumers, such as the stakeholders in academic institutions, can also be maintained and improved. For example, with ISs, they can get updates from their institutions through their email addresses. This can improve efficiency, as they no longer have to come to announcement boards to see for updates, any social activities, or timetable changes.

11.2.3 CHALLENGES OF BIG DATA

There are many challenges faced by organizations and institutions in handling big data. Big data is known to be a group of large data sets and is rich with information that is ready for users to analyze and extract. For many years, big data is still growing massively and until now, leaders are still figuring out how to deal with this particular challenge. This is due to the difficulty to match high demands of big data because the degree of coordination and control are lacking in these areas. Visualization can help in analysis performance and decision making more efficient, however, the issue here concentrates on large volume of data as well as extracting all of the details at high speed [43]. Leaders are trying to solve this challenge by investing in good IS infrastructures to meet objectives and needs by managing big data effectively. Some are resorting towards powerful and increased memory processor to overcome the large amount of data.

Conn [30] mentioned that, the challenges do not only focus mainly on the volume of data, but also from variety and velocity. Thus, to manage big data effectively, leaders must also assess these issues. In the case of an academic institution, data volume problems like overbearing amount of data can come from accumulated of old types of data that have been stored for many years plus the additional new types of data. For example, old and new records or personal details of students that goes into the institution. This increases volume in the storage, which results in too much of data that can be difficult to analyze. Quality of data will also reduce if the data is not accurate or timely. Institutions need to have a quality information management system to ensure that data is always clean. It is best to address issues on data quality and invest on a system, rather than having problems later in time.

In the case of Variety, the problems can come from difficulty to analyze many types of information from data. These data can be extracted through many sources, such as those from tabular data (database), social media, video downloaded from the Internet, images, audio, financial transaction, emails, documents, and many more. With velocity, this involves data streams, availability to access and deliver the data, how quick data is being made, how quick the data is being made and others. Velocity, as previously described, deals with the speed of data streaming. The key issue in velocity is the high requirements of end-users that have streamed data over their personal devices. Data transfers usually do not take much capacity of systems. Transfer rates are limited, but not the requests for them. Therefore, data transfer is the main issue in big data. As for now, the only solution is by shrinking the size of data being sent in these data transfers.

Shrinking size of data can be exampled by Twitter interactions. Most of the interactions present on Twitter are mainly texts, which can be compressed at high rates - easily. In terms of academic, many data available could also be modified in the same case as Twitter, with the fact that they are mainly consisting of texts. Journals, essays, and much paperwork can be compressed to reduce the size of data volume that can then increase the speed of data transfers. Consequently, increasing the velocity in these 3Vs.

Conn [30] also stated that, the real issue of big data is how people utilize it and their initiative to look for patterns that can help institutions make better decisions out of it. Yvonne Genovese mentioned the benefits of implementing Pattern-Based Strategy, emphasizing on how the strategy can help institution by identifying solutions. By implementing the strategy, it can allow users to adapt to changes by identifying opportunity and threats. It also helps to balance diversity of institutional activities such as by defining creativity and collectivity that can enable users to lead and respond to changes of weak and strong signals namely opportunity or threat. The aim of this strategy pattern recognition is to understand different elements that can come from many areas such as activities, events, objects, and information. These elements may form into new patterns that represent an opportunity that can be transformed into innovation or even as a threat that can cause a disruption to the business strategy or operation. She also mentioned that this could also be achieved through other mediums such as from social computing analysis or context-aware computing. By doing this, it can eliminate chances of making shortsighted decisions by discovering useful information and knowledge from Big data through the strategy.

Furthermore, big data information that is readily available is not set for analysis. For instance, a group of students' performance reports in the Ministry of Education, comprising of students from different schools, level of education and such. These data cannot be left in this form to be analyzed. Rather, these data are needed to be extracted in order for it to be suitable for the purpose of analysis. Only required information from the whole bunch of data sets is needed to be extracted for further analysis and doing this correctly and in a complete manner is a technical challenge. It is common to assume the big data is providing correct information when in fact this not always the case. Some people may not input the correct information in the data sets, which can lead to data errors. Existing work on data cleaning assumes that well-recognized constraints on big data or well-understood structure; however, for many well known big data domains, these usually do not exist [23].

11.2.4 WHY BIG DATA IS IMPORTANT TO BE MANAGED

As predicted by Norton and Peel [53], in the last few years, the information management has changed radically with the development and improvement of systems to exploit them fully, resulting in the dramatic transformation of IS (as cited in Ref. [37]). Big data management can be defined as "the organization, administration, and the government of large volumes of both structured and unstructured data" [42]. The aim is to ensure high quality and accessibility for applications purposes. In the academic environment, big data is mostly more important in library usage. Academic libraries are involved with big data in preserving data sets and researching data management. Academic librarians, therefore, have a clear role of managing big data to help improve academic institutions and in bettering the quality of education.

According to Boston University Libraries (n.d.), data can be an important and expensive output of a scholarly research process, across all disciplines. These data are an important part of research results evaluation. They also help to reconstruct events and processes leading to them. The value of data increases as they become aggregated into collections. When they become more available to be reused in addressing new and challenging research questions, the value is unimaginable. This is why data must be managed properly. Without proper data management, value of these data can be greatly diminished.

Big data can be managed through data mining and data analytics. This is to make mining information easier to gain more insights concerning student performance and learning approaches [46]. With data mining, it can make student-related works, such as evaluation, research, and accountability manageable to address. For example, instead of analyzing student's test performance, lecturers can use another approach by finding out the most effective teaching method for each student to improve their performances. Whereas for data analytics, lecturers can use it to focus on different ways of learning. For example, with online tools, lecturers can obtain solutions on a much broader range regarding student actions, how fast they master key concepts, where they usually get their electronic resources and many more [46]. By doing this, lecturers can get the right information to flow to all students in order to improve their productivity and success rates.

Moreover, it is also important to manage big data to help with data analytics. According to Rouse [42], data analytics include the steps required in the inspection of large data sets in order to be able to analyze repetitive models, relationships between data and information and many more. It is required to store and calculate data in a favorable time and precise decisions accurate decisions [27]. The findings from data analytics can be used to improve efficiency of an organization and in bettering customer service, to name a few. Efficiency of an organization can be improved through data analytics by understanding how the organization is working and the efficiency of its performance. Thus, based on the analysis, the organization can strive to make workforce more effective. This would then result in the boost of performance.

Additionally, from a research study carried out at Educause, it was found that in more than half of the institutions responded under their research, 69% reported that data analytics was viewed as an important priority for at least some departments or programs, while 28% regarded analytics as a major priority for the whole learning institution [26]. They also found out that respondents said that data analytics has increased in its importance over the last two years, and will continue to increase in importance in the next few years and in the near future. The respondents also believe that data analytics has a great potential to benefit areas that involve students in an institution.

However, according to Herold [35] with big data analytics, they are not 100% accurate. Data files that are used in the process of data analytics can contain unreliable information—whether these information are about individuals or other things. If the initial data is already incorrect, then the algorithms or results of the data analytics are bad and cannot be used by other people. The risk of having incorrect data can be dangerous as more data added to data sets could mean that more complex data analysis models will be used without the process of validating whether the information is accurate or not. When decisions are made based on these inaccurate and flawed data models, individuals can suffer harm by being denied services or otherwise treated inappropriately.

It is also important to manage big data because it helps with the process of decision-making. Every organization realizes that information is very crucial in helping to decide on a certain matter, academic institutions are no exception. Many institutions use various sources of information for planning, analyzing trends, managing performance and other functions. Consequently, according to Wise [54], the value of information is only as good as its point of entry into the system. The reason for this is because when decision makers analyzed different informations that can lead to making the wrong decisions, which can affect the whole institution, the blame cannot be put on the information itself because the invalidity usually occurs in the process of data entry into the system. Data entry errors and inefficiency in processing information are examples of causes for the prone in data errors that are used to make up an institution's decisions.

11.2.5 MANAGEMENT OF BIG DATA USING CLOUD

According to Hashem et al. [34], cloud computing is defined as a model for allowing ubiquitous, convenient, and on-demand network access to a number of constructed computing resources. These accessibility ranges from networks, server, storage, services, and application. All of these tasks can be supported and supplied by cloud computing with minimal management effort. The cloud is created to handle different kinds of data that can range from external data, internal data, data that comes from personal sources such as personal mobile phone, tablet, personal computer or those data that comes from workers, partners, or even the environment. Organizations, including institutions, rely on this information to carry out everyday activities because it connects the world inside the institution with the world outside the institution.

Managing big data requires high performing IS due to its volume, variety, and velocity nature. Most people usually note on the size of data, but in some cases, it does not only imply on the data volume but also the necessity of scalability. When scalability is considered, solutions like traditional database management systems are already out of context. Cloud computing is a paradigm shift of technology that provides computing services over the Internet. It fits the situation perfectly as it offers benefits such as flexibility in using the computing resources, storage capacity, less management effort, and flexible costs [33]. It composed of highly optimize virtual data centers that have several software, hardware, and information resources that can be used when it is needed by the institution.

Both cloud computing and big data are interrelated because cloud computing provides the basis for big data environment, such as data analysis, accessibility, storage, and easy distribution of information. One advantage of cloud computing is that it is cost-efficient. This is because the services can be deployed without the need of having physical hardwares.

However, although the cloud provides benefits like storage of data as well as data regulating properties, sometimes the flow of data is limited by boundaries that prevent the right data to flow at the right time and at the right place. IBM has come up with a solution to prevent this issue from recurring again. It is known as IBM Cloudant, a database service platform built to ensure flow of data between the application and its database remains uninterrupted, and highly performant (IBM Analytics, n.d). It also ensures that data accessibility for online or offline premises always in smooth sailing without any limitations. The Cloudant also enable users to be connected through interactions. Another service that the cloud provides in the Cloudant is a built-in self-service data refinery called Data works. It filters information analyzed from the data. It is very simple and secure enough to use, even for personal data (IBM, n.d). Organizations or institutions are the perfect candidates to use the cloud because they usually handle a large volume of data that ranges from personal to non-personal ones. For example, records, and details of students and lecturers, or private and confidential data, like financial information of the institution and many more.

Talia (as cited by [34]) claimed that cloud-computing infrastructures could act as an effective platform to address the data storage needed to carry out big data analysis. Through the cloud, raw data can easily be converted into valuable information, which can serve many purposes in making important decisions. Several applications are offered to users to help them manage big data effectively. For example, applications like Hadoop, Amazon Web Services Elastic MapReduce, Google's Big Query, and Big Data Suite.

A virtual warehouse in the cloud called DashDB can be used to store a large amount of data as well as perform analysis (IBM, n.d). Because the warehouse is, cloud-based there is no space limitation. The volume of data in the storage can be expanded and scaled if the users are ever to receive new and greater amounts of data. The cloud can simplify the data and further analyze them to extract more information and resources.

Furthermore, implementing cloud computing can also help manage big data through virtualization. Hashem et al. [34] stated that it is a process that enables sharing resources that reduces the necessity of having physical hardwares around to increase computer resource utilization, scalability, and efficiency. Through visualization, data can be viewed analytically through different graphs, which can aid in decision-making processes. It is also efficient in terms of being cost-efficient with the help of Hadoop clusters, virtualization can help reduce cost by bringing big data analytics more accessible through reducing computer and storage hardwares. Big data can also help users to use commodity computing to process distributed queries across multiple data sets and get the set of results in a timely manner [34]. Cloud computing can also help realize these objective through the use of a software called Hadoop, a platform that enables data to be processed and distributed. It also provides a service that automatically scales demand of users for data processing.

Other than Hadoop, MapReduce is also preferable for large processing of big data in the cloud environment. This software allows large amounts of data sets to be processed and stored parallel in the cluster. Tools present in MapReduce called Hive and Pig makes data processing more feasible to process large data sets easily. Hashem et al. [34] mentioned that cluster computing provides good support to manage data growth within the context of big data.

Additionally, different cloud models manage big data differently: public, private, and hybrid. For example, public cloud offers scalability and elasticity, which is based on a pay-per-use model. Private cloud offers strict control of which is based on on-premises infrastructure. Hybrid cloud, on the other hand, is the mixture of private and public cloud services with a combination between the two. Since the private cloud has a greater security control over data storage, problems like accidental or malicious access can be prevented through shared resources. For public cloud, it encourages visualization and shared of physical resources for data transfers, storage, and processing. Although private clouds provide tighter security than the two cloud models, public cloud is more favorable and suitable to handle a large volume of big data tasks because of its flexibility. However, potential concerns such as bandwidth limitations and costs incurred with data transfer may be at risk.

Hashem et al. [34] also highlighted the potential drawbacks that can result from big data management and selecting cloud-computing methods. The first drawback is lack of data availability; users tend to focus their decisions more on analytical methods, which can be expensive. Moreover, incorrect use of methods or methods that inherent weaknesses can result in making wrong and costly decisions. To address these challenges, a common system called database management system software (DBMSs) is usually used. It plays an important role in ensuring an easy transition of applications from old enterprise infrastructures to new cloud infrastructure architectures [34]. These challenges put pressure on users to focus on settling high big data demands and big data storage, which prompts them to acquire the right technologies - which is by implementing cloud computing.

11.3 DISCUSSION

Big data plays a significant part in plays an important role in academic institutions. From the analysis of the literature review, it can be noted that the purpose of big data in academic is mainly in the circle of e-Learning. Big data has contributed in the rapid progress of the academic environment in terms of online learning. Many systems have been developed to help organize the use of data in higher learning institutions. LMS especially, have been constantly improved in order to keep up with the vastly growing number of data in the web world. Many higher learning organizations have invested a generous amount of money to have systems installed in their organizations. These systems are hoped to help them in organizing data that can then help in many organizational processes, especially in decision-making.

For example, decision-making in terms of arranging schedules for students who are taking many different courses in an institution require having organized information at a high speed to have the task done. This would be such a difficult task to do manually as it will take a longer time to organize everything to fit into one schedule without having clashes. The information needed to analyze these are also large in volume, therefore, a system is needed to increase the efficiency in data entry. By using, a particular system such as the Student Management System mentioned previously, data will be automatically entered and analyzed in the system. The details of institutions can also be kept as records in a system such as the School Management System that keeps details of types of school and its locations as noted by Desilva [32].

This is also the same with lecturers; the system can help in identifying problems like capabilities of lecturers in terms of knowledge of teaching. For example, data in the system can be used to track how good the lecturers are doing in their teaching by looking at the student's progress report. This way, the administration team can think of ways to provide solutions by suggesting different teaching techniques to improve the student's grades. Thus, both IS and big data does not only help in enhancing productivity in students but also to teaching staffs, which is equally important for the institution.

Other than that, another example of this kind of system is the Universities and Colleges Admission Service, also known as UCAS. This is an example of a system that centralizes all information on different universities in the United Kingdom to be easily accessible for international students. This system also helps in registration processes, such as the admission of the students' personal statements and acceptance letters from their respective universities. By having this system, many universities will have more admissions and it would be easier for them to accept or reject an application. This is because accepted students can also have a peace of mind, as their offer letters will certainly be sent to them through emails, and not get lost in the mailbox. They will be notified via a message feature in the system if they have been accepted into a certain university.

Other ISs can also aid in decision-making from the financial perspective. Data volume in big data needs to be managed effectively in order to reduce data redundancy or inconsistency. Conn [30] stated that the increase volume in storage results in too much data that can be difficult to analyze. The quality of data can be reduced if it is not properly managed. This is because data increases in value after it has been analyzed and correlated. As an example, financial information of an institution is usually in big chunks of data. When this financial information is organized and managed properly, institutions can find out where to invest their money to gain advantages from rather than not using the money efficiently. The budget of organization can also be planned and managed effectively when having this kind of system installed. Errors in calculations can be reduced and this can cut costs efficiently. Although the system might be expensive to install, it is good to invest in the system rather than having financial problems in the future. This kind of IS could be given authorization for it to be managed only by certain trusted officers so money laundering or similar kinds of activities will not be an issue.

In higher learning institutions, the department or group, which uses big data the most is their libraries. Libraries of higher learning institutions consist of much information from scholarly articles, digitized textbooks, and journals, graduate research documents and many more. The volume of data in library databases must be constantly updated in order for students to have a more recent insight on their research studies or their current modules. This can help improve students' understanding on certain topics and hence, increasing the quality of their education overall. Many libraries have authorized storage of outsourced information. For example, University Brunei Darussalam has EBSCOHost, Emerald, and SAGE, to help their students in doing their academic research. For libraries to have a good management IS, old data in their databases can be compressed for the purpose of storing records. This is to save the percentage of storage available in their database system and to avoid system breakdown or slow system performance. This is the reason why the scalability of databases is an important subject to consider in managing data.

The scalability of the IS - is an essential feature of managing big data. As the volume of data increases, the computer system must be capable to manage this high volume of data. With every passing year, data entered in the system will certainly increase. With every increase in the usage of data, it also increases the workload of database performance in terms of processing and managing data. If the system is not scalable, it can cause a lot of issues, such as slow performance or down system problems because the system is not equipped to handle heavy load tasks. This was confirmed by Fernandez et al. [33] where traditional database will be out of big data context as the system is not capable to handle big data requirements. However, investing in scalable IS to handle big data ensures full optimization of productivity because it helps the institution to react to new conditions without having to face system failure. As a result, to this, it is important for the system to have a robust technology that can automatically recover whenever these kinds of problems occur.

Additionally, in order to be efficient in cost-reduction, institutions need to have systems that are high in scalability, which can meet the needs of both current and future data requirements. If the systems are not scalable, they are usually difficult to be repaired and it may take up days to complete which can result in work progress being hindered or in worse case scenarios, they might have to purchase a new database management system every now and then, which is a waste in terms of finance and time. For example, an institution usually implement a system called geographic information system (GIS) that is loaded with a lot of information regarding module registrations, module timetables, examination results, progress report, and many others. If the system is not performing well, GIS can experience breakdown, which can cause difficulty for students and lecturers to view information on the system.

Other than that, distribution of information will be delayed which can create further problems. For example, during module registrations, students will have to compete with one another in the attempt to get into their chosen modules due to the maximum quota that is already assigned to each class. If the system is slow, this can reduce the chances of students being accepted to each module, which is very inefficient. Furthermore, accuracy of the data being delivered is also equally important. This is in terms of allocation of timetables. Students and lecturers depend highly on this information to be on track in order for them to be able to divide and manage their time well. If this information is miscommunicated or there are errors in between, it can disrupt lecturers and students' activities as well as the level of their productivity. Therefore, it is important for an institution to test and do research on the systems they want to implement in order to get better results out of them instead of adding more incapable systems. This shows that scalability is an important aspect of big data because it promotes efficiency in terms of conveying information to respective stakeholders as well as helping the institution to be cost-efficient.

The process of data analytics that was mentioned by Herold [35] is also an important issue that needs to be addressed. In academic institutions, having inaccurate information is not acceptable as it may lead to unfortunate consequences. Students acquire much information from the Internet and other web-based sources as input for their studies. If there is much incorrect information made available on big data, it can create a haywire in the education system. Mistakes in analyzing data can happen in data analytics of academic institutions. For example, students who are tagged with the wrong student identification number can be entered into the wrong classes and this can result in students taking the wrong examination papers.

The use of cloud in academic has also increased in importance ever since it was first introduced. Many higher learning institutions are now investing in cloud systems. This is because the traditional way of keeping information is now considered as costly and ineffective. Paperless system is more in favor nowadays as the academic environment has also taken part in improving the status of environmental issues. By using cloud systems, many financial costs are being cut and this also promotes a healthier and greener way of running administrations. In going hand in hand with preserving natural resources of the environment, many institutions get marketed better as it can mean that they are morally ethical than those who do not practice 'going green.' This can also have an effect in student admissions and ranking systems of universities.

As the volume of data increases, the need for increase in velocity is also there. Therefore, cloud systems can help speed up organizational processes in such a way by using the Internet. Much of the information stored in the clouds can be accessible by many stakeholders in the academic field all at the same time. If institutions do not make use of these cloud systems, other institutions can leave them behind. Institutions, which do not use cloud systems have to do their processes the traditional way (manually) which can be timely. Competitive advantage can be achieved if institutions have a more updated system that is more scalable and also, more cost-efficient. High volumes of data can be analyzed and processed at a higher speed. This can then improve the performance of an institution as a whole.

Cloud systems can be at use for both educators and learners. For example, students can do their assignments on the go while using cloud systems. It would not matter if they do not own or have their personal computers with them, they can still access the same information from the cloud. The same goes to educators or lecturers. Computers that are provided by their universities can have the same files as their personal ones at home, providing they do their work on the cloud. As stated by Hashem et al. [34], the use of commodity computing can distribute queries across multiple data sets and get results in a timely manner.

Although there are many beneficial impacts that can be gained from big data in academic, many issues have also arisen in the process of implementing it and the systems that come with it. Firstly, it was mentioned by Marcella and Knox [37], not all stakeholders in academic institutions are in favor of using computers. This can hinder the progress of an institution. Additionally, when there is resistance in the institution, the implementation of a certain system is usually unsuccessful. As a consequence, financial investments made to implement the systems will not be fruitful. On the other hand, if there is no resistance, training is also needed for staff and administrators to acquire the use of newly implemented systems. This can increase the expenses of institutions.

Furthermore, the strategies in choosing the right type of cloud system to be implemented in a certain institution are also a challenge. This is a vital process as implementation and installation of the wrong cloud system can be costly. Academic institutions might be more suitable in implementing a private cloud for its administration as it provides a tighter security than the other two cloud models. However, public cloud can also be good as it is more suitable in handling a large volume of big data tasks.

11.4 CONCLUSION

To conclude the whole discussion, big data in academic is a very familiar topic. It has been present from a very long time and progresses have been made ever since. As the number of big data increases, more ISs have also been developed in managing this huge amount of data available. Big data offers many benefits academically. To summarize, it can improve efficiency, effectiveness, and productivity in academic institutions. All stakeholders in academic institutions can also make use of the big data available. Many academic achievements can be credited to the availability of big data in the academic environment. Systems and application softwares that have helped to analyze big data and convert them into useful information are also another subject that can be appreciated.

Without having big data in academic, many organizational processes in academic institutions would have progressed at a slower rate. This would slow down the development of academic achievements of many institutions around the world. Big data has helped improve academic institutions by providing much useful information. The richer the content of information, more insights can be taken into account. This is the reason to why big data needed to be mined, processed, analyzed, or compressed into aggregated information so that it can be useful to academic institutions.

Furthermore, the availability of cloud computing systems has taken the storage of data to a whole other level. These systems are able to have allowed for outdated information to still be kept in storage. These old information can be of help in analyzing future trends in the academic environment and not necessarily be a waste of space or data storage. Therefore, a higher number of data can be stored more efficiently. Current technologies are also constantly improving in the managing of big data's 3Vs. The volume, velocity, and the variety of big data will always be something that needs to have improvement from time to time.

ISs that have been implemented in academic institutions have also contributed a lot in the development of academic progress. Without the application of these ISs, many processes in the institutions would have taken more time and used more of the institutions' financial budget. With the increasing amount of big data in every passing year, academic institutions would have to constantly prepare for installation of new databases in case the current ones being used are no longer suitable with the needs of the institution. Scalability of databases or system softwares must always be a priority to consider when installing new ones.

Having proper planning in data management, on the other hand, can reduce the challenges of big data in academics. With the help of data management and data analytics, large data sets can be used to improve academic institutions as a whole. Data used to analyze patterns and trends can be stored and computed to help make accurate decisions for institutions. Efficiency of academic institutions can be improved through data analytics as performance of the institutions is continuously monitored. Consequently, institutions can be more effective and the performance will consistently be improved.

Every set of big data is considered as having a great significance. Each and every information is or can be important in the process of decision-making. This is another reason why it is vital for academic institutions to ensure that there IS - is on point and constantly updated. Many institutions sometimes have technical problems occurring in their systems, resulting in the deterrence of their organizational processes. This affects the productivity of many academic institutions. This problem needs to be tackled quickly as to reduce the chances of institutions to be at a cost disadvantage. Performance hindrance can be expensive if it is not fixed. All ISs used in academic institutions must therefore be maintained properly to have a constant excellent performance in its systems.

Moreover, many budgets of academic institutions have been allocated in the investment on ISs to handle and manage big data. Although these investments are very costly, it can be approved that it is best to invest on these than to miss out on opportunities that can be grabbed by these institutions. Opportunities in and for institutional development can be initiated by analyzing information from big data. Therefore, information is needed to be managed constantly, and organized efficiently. Data entry errors to systems need to be reduced in order to improve the productivity of academic institutions. Training for technical staff in handling ISs is vital as they are the ones who would be responsible if there is to be any error found in the system. Like data entry errors, individuals, and not the system itself cause much of other blunders that occurred in many ISs.

All stakeholders in academic institutions have benefited from big data, especially lecturers and students. Lecturers can improve on teaching materials and have the ability to select teaching styles, which are best fit and students can get better grades from understanding more of their academicals input. With the help of big data in academic, many undergraduate students are able get their research done by analyzing web-based data sets. Comparisons of articles and journals have been made easy as many of them have been digitized and stored on the Internet. Cloud computing has helped in changing the way students work on their assignments. Although libraries are still important, many students have also starting to choose to browse for online materials rather than getting physical books from their institutions' libraries.

All in all, it can be concluded that big data in academic has rapidly improved and will continuously increase the rate of progression in academic institutions. It is hoped that in the future, big data in academic will remain its importance and value as it is today in the current state of academic environment, if not better and in a more advanced state.

11.5 RECOMMENDATION

As mentioned above, big data and IS are both important mechanisms in decision-making process for the institution as a whole. However, there are also challenges faced by the stakeholders that may divert their objectives and hinder them from meeting their goals. Several approaches can be made to eliminate possibilities of making the same errors or new ones in the future. According to Roe [41], planning, defining strategies, and maximizing existing or readily available data can avoid making potential mistakes and wrong decisions in terms of analyzing big data. By planning and defining the right strategy, the institution will have a clear objective of what they should focus on instead of wasting their time on analyzing data that can result to meaningless outcome. He mentioned that by implementing analytics strategy, it will enable the institution to get to know their stakeholders better by understanding their needs. This is because nowadays, more people are moving towards technological advances and are starting to take a huge interest on it. Thus, the need for high performing ISs is increasing. By having, the right ISs, big data can be regulated efficiently and interactions between stakeholders will also be at ease.

The concept of analytic strategy focuses on being strategic that can support in decision-making process like addressing problems and questions arise from it. The process involves by starting with a strategic question, finding or collecting relevant data that can help answer the question, analyzing the data as well as making predictions and gaining insight, illustrating findings that can be both understandable and actionable, and lastly feeding the findings back into the initial process to address the strategic questions and create new ones [26]. He mentioned that, most institution are moving towards this strategy because it is different from a lot of traditional analysis and reporting approaches. What separates this strategy from the old approach is data obtained from the system is more extensive and automatic and the processes used to extract and analyze the data are becoming repeatable [26]. This means that information obtained from the analyzed data can always be used several times whenever the institution needs it. Therefore, because of this, Bichsel [26] claimed that, the numbers of departments and programs slowly increases in incorporating data and analysis strategy into their decision making and planning process due to the positive feedback.

However, the institution must also ensure that the strategy is able to cover all-important aspects such as the overall vision and requirements within the institutional context. When this is done, the institution will be able to predict key challenges that might happen in the future as well as identifying solutions to overcome them. Furthermore, the institution can also discover the process requirements that can help to define how big data can be used including which IS infrastructure, tools, software, and many others before officially implementing the strategy.

Moreover, before planning and implementing the strategy, it is crucial for the institution to know if their objectives can be achieved realistically. This can be in terms of whether the strategy can give value to the institution rather than guiding it in the wrong direction. Roe [41] suggested that the right way to do this is to utilize and maximize the data that is readily available in the institution. This includes utilizing existing software and information surrounding the institution. By doing this, the institution will be able to expand and analyze the information better. Existing software can be modified according to the needs and requirements of the institution to enhance scalability and effectiveness.

Most importantly, the key to manage big data efficiently is to have the right IS. Institution must be able to keep them informed with new and updated technologies as well as softwares in order to know which one is more fitting to meet the requirements of their stakeholders. This is to provide better tools to manage the constantly increasing big data and to be able to extract maximum amount of information as possible in order to gain valuable information from it. Other than that, having the right people with the right set of skills is also equally important. Bichsel [26] emphasizes on the importance of investing in a number of analyst before investing in IS tools. These people can serve as a backbone to support major problems like system errors and down system problems that can happen in the future which can deter process of data analytics. With their knowledge, systems will be restored back to its initial condition. The analyst can also be of help in defining strategies together and ways of solutions to address them. This will not only promote productivity but will also stimulate pool of ideas to help the institution to perform better. Therefore, surrounding the institution with the right skill sets as well as ISs will not only help in big data management but it will also be a good investment which will certainly reward the institution in many aspects.

For example, there are many institutions, which have benefitted from following these steps. According to Bichsel [26], he found out that, most claimed this strategy have helped them in several departments in the institution in terms of how they utilize the data in several functional areas. From the results obtained, many institutions, which have implemented this strategy, manage to spread positive results to all of its stakeholders. In implementing the analytics strategy, most institutions also introduced analytic programs that encourage the stakeholders to participate in. Students and lecturers, who have participated, claimed that data analytics help to foster their interactions, communication, and decision-making as well as boost up their morale in class. For example, it can help with students' enrollment, management, budgeting, finance, and student progress.

Others mentioned how the analytic programs are very systematic in increasing the likelihood of faculty, staff, and administrators to base their decisions on data rather than on intuition or conventional wisdom [26]. This instilled a form of culture where the stakeholders will treat big data as high importance. Some leaders often follow their intuition or experience or even their preconceived ideas to influence the decision-making process which can reduce the quality of the outcome. This also wastes time and money if they are to implement the wrong strategy. Thus, it can be suggested that analytics strategy helps the academic community to center their attention on strategically important questions only which help to encourage continuous improvement within the institution and amongst its stakeholders.

He also found out that the strategy could help in increasing satisfaction between the students, teaching staff, administrators, other faculty staff, and many others. It helps to reduce political conflict between the stakeholders because decisions are no longer biased but will be based on data that have been extracted and analyzed. This can lead to more factual and concrete outcomes rather than wasting time and money in extracting and collecting additional data that does not lead to anything. This made the whole process of decision making more effective, efficient, and improved, which is good for the long-term basis of the institution.

Other than helping in decision-making areas, the strategy can also help in enhancing students' progress and their performance as a whole. The strategy can help in extracting data for grade information, demographics, and student personal details, existing, and new student data on effort and engagement for each subject they take. This helps to keep track of their early performance as well as study behaviors until the end. When their performance is decreasing in between the semesters, the person in charge will be notified. Steps and different approaches can be taken to improve this issue by consulting the students regarding their overall performance. This approach can result in high grades in students because it can help them be informed of what went wrong with their study and learning methods and how they should tackle the issue. The good part about this is that the academic community can identify students who need additional help that can make the overall performance of the institution better.

Apart from this, the institution must also include the importance of data access by stakeholders in the strategy. The problem of data access is always highlighted because institutions sometimes neglect this aspect which can create more problems amongst the stakeholders such as miscommunication of information, failure to distribute information or distributing information to the wrong person. To minimize this error, the DataMaster system can be used to create data that is isolated which can be accessible and centralized, data silos and also promotes a culture of decision making based on data [26]. Although, Bichsel [26] argued that it is impossible to make an isolated for the whole institution, other measures can be taken to ensure data is accessible to everyone. The institution must first need to have a consistent theme, which is a common language around the data so that data can be centralized to make it more accessible and usable. This will ensure data consistency and can be used by various departments. When data is consistent and centrally accessible, it can serve as a better foundation for different variations of decision-making groups.

KEYWORDS

- **big data**
- **database management system software**
- **engineering systems**
- **geographic information system**
- **information system**
- **virtual learning environments**

REFERENCES

1. Susanto, H., & Almunawar, M. N., (2018). *Information Security Management Systems: A Novel Framework and Software as a Tool for Compliance with Information Security Standard*. CRC Press.
2. Susanto, H., & Chen, C. K., (2018). Macromolecules visualization through bioinformatics: An emerging tool of informatics. *Applied Physical Chemistry with Multidisciplinary Approaches, 383*.
3. susanto, H., & Chen, C. K., (2018). Informatics approach and its impact for bioscience: Making sense of innovation. *Applied Physical Chemistry with Multidisciplinary Approaches, 407*.
4. Susanto, H., (2000). Smart mobile device emerging technologies: An enabler to health monitoring system. *Kalman Filtering Techniques for Radar Tracking, 241*.
5. Liu, J. C., Leu, F. Y., Lin, G. L., & Susanto, H., (2018). An MFCC-based text-independent speaker identification system for access control. *Concurrency and Computation: Practice and Experience*, *30*(2), e4255.
6. Almunawar, M. N., Anshari, M., Susanto, H., & Chen, C. K., (2018). How people choose and use their smartphones. In: *Management Strategies and Technology Fluidity in the Asian Business Sector* (pp. 235–252). IGI Global.
7. Susanto, H., Chen, C. K., & Almunawar, M. N., (2018). Revealing big data emerging technology as enabler of LMS technologies transferability. In: *Internet of Things and Big Data Analytics Toward Next-Generation Intelligence* (pp. 123–145). Springer, Cham.
8. Almunawar, M. N., Anshari, M., & Susanto, H., (2018). Adopting open source software in smartphone manufacturers' open innovation strategy. In: *Encyclopedia of Information Science and Technology,* (4th edn., pp. 7369–7381). IGI Global.
9. Susanto, H., (2017). Cheminformatics—the promising future: Managing change of approach through ICT emerging technology. *Applied Chemistry and Chemical Engineering: Principles, Methodology, and Evaluation Methods*, *2*, p. 313.
10. Susanto, H., (2017). Biochemistry apps as enabler of compound and DNA computational: Next-generation computing technology. *Applied Chemistry and Chemical Engineering: Experimental Techniques and Methodical Developments*, *4*, 181.
11. Susanto, H., (2017). Electronic health system: Sensors emerging and intelligent technology approach. In: *Smart Sensors Networks* (pp. 189–203).
12. Leu, F. Y., Ko, C. Y., Lin, Y. C., Susanto, H., & Yu, H. C., (2017). Fall detection and motion classification by using decision tree on mobile phone. In: *Smart Sensors Networks* (pp. 205–237).
13. Susanto, H., & Chen, C. K., (2017). Information and communication emerging technology: Making sense of healthcare innovation. In: *Internet of Things and Big Data Technologies for Next Generation Healthcare* (pp. 229–250). Springer, Cham.
14. Susanto, H., Almunawar, M. N., Leu, F. Y., & Chen, C. K., (2016). Android vs. iOS or others? SMD-OS security issues: Generation Y perception. *International Journal of Technology Diffusion (IJTD)*, *7*(2), 1–18.

15. Susanto, H., (2016). *Managing the Role of IT and IS for Supporting Business Process Reengineering.*
16. Susanto, H., Kang, C., & Leu, F., (2016). *Revealing the Role of ICT for Business Core Redesign.*
17. Susanto, H., & Almunawar, M. N., (2016). Security and privacy issues in cloud-based e-government. In: *Cloud Computing Technologies for Connected Government* (pp. 292–321). IGI Global.
18. Leu, F. Y., Liu, C. Y., Liu, J. C., Jiang, F. C., & Susanto, H., (2015). S-PMIPv6: An intra-LMA model for IPv6 mobility. *Journal of Network and Computer Applications, 58*, 180–191.
19. Susanto, H., & Almunawar, M. N., (2015). Managing compliance with an information security management standard. In: *Encyclopedia of Information Science and Technology*, (3rd edn., pp. 1452–1463). IGI Global.
20. Almunawar, M. N., Susanto, H., & Anshari, M., (2015). The impact of open source software on smartphones industry. In: *Encyclopedia of Information Science and Technology,* (3rd edn., pp. 5767–5776). IGI Global.
21. Almunawar, M. N., Anshari, M., & Susanto, H., (2013). Crafting strategies for sustainability: How travel agents should react in facing a disintermediation. *Operational Research, 13*(3), 317–342.
22. Nabil, A. M., Susanto, H., & Anshari, M., (2013). A cultural transferability on IT business application: I-reservation system. *Journal of Hospitality and Tourism Technology, 4*(2), 155–176.
23. Agrawal, et al., (2012). *Challenges and Opportunities with Big Data: A White Paper Prepared for the Computing Community Consortium Committee of the Computing Research Association.* Retrieved from: http://cra.org/ccc/wp-content/uploads/sites/2/2015/05/bigdatawhitepaper.pdf (Accessed on 2 January 2020).
24. Allen, D., (1995). *Information Systems Strategy Formation in Higher Education Institution.* Retrieved from: http://www.informationr.net/ir/1-1/paper3.html (Accessed on 2 January 2020).
25. Arabasz, P., & Baker, M. B., (2003). *Evolving Campus Support Models for E-Learning Courses.* Retrieved from: http://www.educause.edu/ir/library/pdf/ERS0303/ekf0303.pdf (Accessed on 2 January 2020).
26. Bichsel, J., (2012). *Analytics in Higher Education.* Retrieved from: https://www.researchgate.net/publication/281111069_Analytics_in_Higher_Education_Benefits_Barriers_Progress_and_Recommendations/link/55d6221f08aec156b9a84c48/download (Accessed on 20 January 2020).
27. Big Data Analytics News.com. (2015). *Importance of Big Data Analytics for Business Growth.* Retrieved from: http://bigdataanalyticsnews.com/importance-of-big-data-analytics-for-business-growth/ (Accessed on 2 January 2020).
28. Boston University Libraries, (n.d.). *Importance of Data Management.* Retrieved from: https://www.bu.edu/dioa/data-management/ (Accessed on 20 January 2020).
29. Clayton-Pedersen, A., & O'Neill, N., (2005). *Educating the Net Generation.* Retrieved from: https://www.educause.edu/ir/library/PDF/pub7101.PDF (Accessed on 20 January 2020).

30. Conn, S., (2011). *Gartner Says Solving 'Big Data' Challenge Involves More Than Just Managing Volumes of Data*. Retrieved from: http://www.gartner.com/newsroom/id/1731916 (Accessed on 2 January 2020).
31. Costello, E., (2016). *Opening up to Open Source: Looking at how Moodle is Adopted in Higher Institution*. Retrieved from: http://flosshub.org/sites/flosshub.org/files/Moodle-Adoption-in-Higher-Education-Eamon_Costello.pdf (Accessed on 2 January 2020).
32. De Silva, S., (2015). *The Impact of Education Management Information Systems: The Case of Afghanistan*. Retrieved from: http://blogs.worldbank.org/education/impact-education-management-information-systems-case-afghanistan (Accessed on 2 January 2020).
33. Fernandez, et al., (2014). *Big Data with Cloud Computing: An Insight on the Computing Environment, MapReduce, and Programming Frameworks*. Retrieved from: http://sci2s.ugr.es/sites/default/files/ficherosPublicaciones/1810_2014-WIRES-Fernandez_etAl-Big_Data_w_Cloud_Computing.pdf (Accessed on 2 January 2020).
34. Hashem, et al., (2014). *The Rise of "Big Data" on Cloud Computing: Review and Open Research Issues*. Retrieved from: http://www.sciencedirect.com/science/article/pii/S0306437914001288 (Accessed on 2 January 2020).
35. Herold, R., (2016). *10 Big Data Analytics Privacy Problem*. Retrieved from: https://www.secureworldexpo.com/10-big-data-analytics-privacy-problems (Accessed on 2 January 2020).
36. Kotadia, H., (2016). *Five Ways Big Data are Fundamentally Changing the Information System*. Retrieved from: http://customerthink.com/5_ways_big_data_are_fundamentally_changing_information_systems/ (Accessed on 20 January 2020).
37. Marcella, R., & Knox, K., (2004). *Systems for the Management of Information in a University Context*. Retrieved from: http://www.informationr.net/ir/9-2/paper172.html (Accessed on 2 January 2020).
38. Nichols, M., (2003). A theory of eLearning. *Educational Technology and Society, 6*(2), 1–10.
39. Oblinger, D. G., & Oblinger, J. L., (2005). *Educating the Net Generation*. Educause. Retrieved from: https://www.ds.unipi.gr/et&s/journals/6_2/1.pdf (Accessed on 20 January 2020).
40. Rand, N., (2014). *Five Quick Links: Managing Big Data in the Cloud*. Retrieved from: http://searchcloudcomputing.techtarget.com/feature/Five-quick-links-Managing-big-data-in-the-cloud (Accessed on 2 January 2020).
41. Roe, D., (2013). *Five Recommendations for Developing a Big Data Analytics Strategy*. Retrieved from: http://www.cmswire.com/cms/information-management/5-recommendations-for-developing-a-big-data-analytics-strategy-019162.php (Accessed on 2 January 2020).
42. Rouse, M., (2016). *Big Data Management*. Retrieved from: http://searchdatamanagement.techtarget.com/definition/big-data-management (Accessed on 2 January 2020).
43. Sas.com. (2013). *Five Big Data Challenges*. Retrieved from: http://4instance.mobi/16thCongress/five-big-data-challenges-106263.pdf (Accessed on 20 January 2020).

44. Sharma, A., & Vatta S., (2013). *Role of Learning Management Systems in Education.* Retrieved from: https://pdfs.semanticscholar.org/3d19/dc963e7fb8ce49bb6bcc9329aa03e22a6075.pdf?_ga=2.234289115.457661162.1579679839-2086552453.1579679839 (Accessed on 20 January 2020).
45. Syafique, F., & Mahmood, K., (2010). *The Role of Educational Information Systems for Survival in Information Society and the Case of Pakistan.* Retrieved from: http://www.academia.edu/1190508/The_role_of_educational_information_systems_for_survival_in_information_society_and_the_case_of_Pakistan (Accessed on 2 January 2020).
46. West, D. M., (2012). *Big Data for Education: Data Mining, Data Analytics, and Web Dashboards.* Retrieved from: https://www.brookings.edu/research/big-data-for-education-data-mining-data-analytics-and-web-dashboards/ (Accessed on 2 January 2020).
47. Zwass, V., (2016). *Information System.* Retrieved from: https://www.britannica.com/topic/information-system (Accessed on 2 January 2020).
48. Harasim, L. (2000). Shift happens: Online education as a new paradigm in learning. The Internet and higher education, *3*(1–2), 41–61.
49. Kerry, A., Ellis, R., & Bull, S. (2008, December). Conversational agents in E-Learning. In *International Conference on Innovative Techniques and Applications of Artificial Intelligence* (pp. 169–182). Springer, London.
50. Ebardo, R. A., & Valderama, A. M. C. (2009, December). The effect of web-based learning management system on knowledge acquisition of information technology students at Jose Rizal University. In *Proceeding of 6th International Conference on E-learning for Knowledge-Based Society, Bangkok, Thailand.*
51. Samsonov, P., & Beard, M. (2005). Implementing Blackboard Online Delivery System at a High School: Lessons learned. In *Society for Information Technology & Teacher Education International Conference* (pp. 1578–1582). Association for the Advancement of Computing in Education (AACE).
52. Dwyer, K., & Holte, R. (2007, September). Decision tree instability and active learning. In *European Conference on Machine Learning* (pp. 128–139). Springer, Berlin, Heidelberg.
53. Norton, B., & Peel, M. (1989). Information: The key to effective management. *Library Management.*
54. Wise, A. F., & Shaffer, D. W. (2015). Why theory matters more than ever in the age of big data. *Journal of Learning Analytics, 2*(2), 5–13.

INDEX

F

G

H

I

N

O